Solving Statics Problems in Mathcad

Brian D. Harper
Ohio State University

ENGINEERING MECHANICS
STATICS

Sixth Edition

T0225835

J. L. Meriam
L. G. Kraige
*Virginia Polytechnic Institute
and State University*

John Wiley & Sons, Inc.

CONTENTS

INTRODUCTION

Computers and software have had a tremendous impact upon engineering education over the past several years and most engineering schools now incorporate computational software such as Mathcad in their curriculum. Since you have this supplement the chances are pretty good that you are already aware of this and will be having to learn to use Mathcad as part of a Statics course. The purpose of this supplement is to help you do just that.

There seems to be some disagreement among engineering educators regarding how computers should be used in an engineering course such as Statics. I will use this as an opportunity to give my own philosophy along with a little advice. In trying to master the fundamentals of Statics there is no substitute hard work. The old fashioned taking of pencil to paper, drawing free body diagrams and struggling with equilibrium equations, etc. is still essential to grasping the fundamentals of mechanics. A sophisticated computational program is not going to help you to understand the fundamentals. For this reason, my advice is to use the computer only when required to do so. Most of your homework can and should be done without a computer. A possible exception might be using Mathcad's symbolic algebra capabilities to check some messy calculations.

The problems in this booklet are based upon problems taken from your text. The problems are slightly modified since most of the problems in your book do not require a computer for the reasons discussed in the last paragraph. One of the most important uses of the computer in studying Mechanics is the convenience and relative simplicity of conducting parametric studies. A parametric study seeks to understand the effect of one or more variables (parameters) upon a general solution. This is in contrast to a typical homework problem where you generally want to find one solution to a problem under some specified conditions. For example, in a typical homework problem you might be asked to find the reactions at the supports of a structure with a concentrated force of magnitude 200 lb that is oriented at an angle of 30 degrees from the horizontal. In a parametric study of the same problem you might typically find the reactions as a function of two parameters, the magnitude of the force and its orientation. You might then be asked to plot the reactions as a function of the magnitude of the force for several different orientations. A plot of this type is very beneficial in visualizing the general solution to a problem over a broad range of variables as opposed to a single case.

As you will see, it is not uncommon to find Mechanics problems that yield equations that cannot be solved exactly. These problems require a numerical approach that is greatly simplified by computational software such as Mathcad. Although numerical solutions are extremely easy to obtain in Mathcad this is still the method of last resort. Chapter 1 will illustrate several methods for obtaining symbolic (exact) solutions to problems. These methods should always be tried first. Only when these fail should you generate a numerical approximation.

Many students encounter some difficulties the first time they try to use a computer as an aid to solving a problem. In many cases they are expecting that they have to do something fundamentally different. It is very important to understand that there is no fundamental difference in the way that you would formulate computer problems as opposed to a regular homework problem. Each problem in this booklet has a problem formulation section prior to the solution. As you work through the problems be sure to note that there is nothing peculiar about the way the problems are formulated. You will see free-body diagrams, equilibrium equations etc. just like you would normally write. The main difference is that most of the problems will be parametric studies as discussed above. In a parametric study you will have at least one and possibly more parameters or variables that are left undefined during the formulation. For example, you might have a general angle θ as opposed to a specific angle of 20°. If it helps, you can "pretend" that the variable is some specific number while you are formulating a problem.

This supplement has seven chapters. The first chapter contains a brief introduction to Mathcad. If you already have some familiarity with Mathcad you can skip this chapter. Although the first chapter is relatively brief it does introduce all the methods that will be used later in the book and assumes no prior knowledge of Mathcad. Chapters 2 through 7 contain computer problems taken from chapters 2 through 7 of your textbook. Thus, if you would like to see some computer problems related to friction you can look at the problems in chapter 6 of this supplement. Each chapter will have a short introduction that summarizes the types of problems and computational methods used. This would be the ideal place to look if you are interested in finding examples of how to use specific functions, operations etc.

This supplement uses Mathcad version 13.0. Mathcad is a registered trademark of MathSoft, Inc., 101 Main Street, Cambridge, Massachusetts, 02142.

AN INTRODUCTION TO MATHCAD

1

This chapter provides an introduction to the Mathcad programming language. Although the chapter is introductory in nature it will cover everything needed to solve the computer problems in this booklet.

1.1 Numerical Calculations

Mathcad has four different equals signs. The most important of these are the *evaluation* equals sign (=) and the *assignment* equals sign (:=). Numerical calculations use the evaluation equals sign. As a simple example, type the following expression into a Mathcad worksheet: "(2+6^3)*4/5=". After pressing the "=" key, Mathcad will immediately evaluate the expression. It should look like the following.

$$\left(2 + 6^3\right) \cdot \frac{4}{5} = 174.4$$

Note that the result looks very much like what you would write on a sheet of paper. Now try typing "10+12/3-6*2^4=" into the worksheet. You should get the following.

$$10 + \frac{12}{3 - 6 \cdot 2^4} = 9.871$$

At first it may surprising that the 6*2^4 remains in the denominator. Now try entering the same keystrokes but press the space bar immediately after typing "3". Note how the blue placeholder changes when the space bar is pressed. With a little practice, you shouldn't have too much trouble getting the expression you want. The main thing is to pay attention to the placeholder. The arrow keys can also be used to move the placeholder.

Numerical calculations can also include standard functions. The most commonly used functions can be found in the calculator toolbar. The calculator toolbar can be opened with *View...Toolbars* or by pressing shortcut button that looks like a calculator. Mathcad has many built in functions besides those shown in the

Calculator toolbar. If you already know the name of the function you can simply type it in or select from a list by using the shortcut Cntrl+E or by choosing *Insert...Function* in the menu bar. Here are a few examples. Explanations are given to the right when appropriate.

$$\sqrt{6 \cdot \frac{4}{18}} = 1.155$$

> Press the square root button in the Calculator toolbar then type "6*4/18="

$$\sinh(0.5) = 0.521$$

> The hyperbolic sine. Either type "sinh(0.5)=" or select *Insert...Function...Hyperbolic...sinh* in the menu bar.

$$\sin(10) = -0.544$$

> Type "sin(10)=" or select sin from the Calculator toolbar.

Mathcad, like most mathematical software packages, assumes that angles are given in radians. Thus the last line calculates the sine of 10 radians (573 degrees). Use one of the following to methods to obtain the sine of 10 degrees.

$$\sin\left(10 \cdot \frac{\pi}{180}\right) = 0.174 \qquad \sin(10 \cdot \deg) = 0.174$$

Of course, inverse trig functions also return results in radians and similar methods can be used to obtain results in degrees. The following calculates an inverse sine (asin in Mathcad) and converts the result to degrees.

$$\frac{180}{\pi} \cdot \operatorname{asin}\left(\frac{\sqrt{3}}{2}\right) = 60 \qquad \frac{\operatorname{asin}\left(\dfrac{\sqrt{3}}{2}\right)}{\deg} = 60$$

1.2 Variables and Functions

A variable is a name or alias which can be defined as a number or an expression using the *assignment* equals sign ":=" (type ":" in Mathcad). Mathcad has many built in variables. A good example is the variable *deg* (an alias for the number $\pi/180$) used in the previous examples. To see this, type "deg=" in a Mathcad worksheet. Of course, you can also define your own variables and functions in Mathcad. The following example assigns a number to the variable x and an expression (a function of x) to the variable f. Technically, both x and f are variables though it is customary to refer to f as a function of x. Following the two assignments we also use the evaluation equals sign (=) in order to illustrate the difference between these two equals signs. As the names suggest, one is used for assigning (giving names to) numbers or expressions while the other is used for evaluating (calculating) names or expressions.

$x := 5$ | Type "x:5" |

$f := 3 \cdot x - 5 \cdot x^2 + 2 \cdot x^3$

$f = 140$

Assigning expressions to names is very useful when you want to calculate the values of a function for several different values of a parameter. Note, however, that x must be assigned a numerical value before assigning the expression above to the name f. It is also possible to define functions explicitly in terms of one or more parameters. In this way you can define functions that work just like built in functions such as sin, cos, log etc. When functions are defined in this way it is not necessary to specify beforehand the values of the parameters in the equation. Here are a few examples.

$f(y) := 3 \cdot y - 5 \cdot y^2 + 2 \cdot y^3$ | to enter a function type "f(y):" followed by the expression |

$f(5) = 140 \quad f(2) = 2$ | note that the function f operates like a built in function |

$g(x,y) := \sqrt{x^2 + y^2}$ | note that it is okay to use x as a parameter in a function definition even though it has been previously defined a value |

$g(x,x) = 7.071$

> note the evaluation equals sign. Now Mathcad substitutes the value previously assigned to x (5) into the function g, resulting in the square root of 5^2+5^2

$g(4, f(2)) = 4.472$

> One function can be used as the argument for another. Mathcad first evaluates f(2) and then substitutes this for y in the function g(x,y).

Range Variables

As the name implies, a range variable is a variable which has been assigned a range of values. Assigning a range to a variable is accomplished by typing something like "x:a,b;c" where a, b and c are numbers or variables previously assigned a numerical value. The first value in the range for the variable x is a, the second value is b while the last value in the range is c. Note that b is the second **value**, not the increment. Mathcad will automatically determine the increment from a and b. Let's try it out. Type "x:1,1.5;3" followed by "x=". You should see the following.

$x := 1, 1.5 .. 3$ $x =$

1
1.5
2
2.5
3

Notice that two dots (..) are displayed when you type the semicolon (;). The two dots is Mathcad's *range variable* operator. A shortcut (m..n) is available on the Calculator toolbar. Once a range variable has been defined it can be used like any other variable.

$z := 0, -1 .. -6$

> Type "z:0,-1;-6". Notice that the range can either increase or decrease!

$f(y) := 2 \cdot y + y^3$

> Type "f(y):2*y+y^3"

$f(3) = 33$

$f(z) =$

0
-3
-12
-33
-72
-135
-228

Type "f(z)=". Notice that the function f can operate either on a single value or a range of values.

If the second value in the range is omitted, Mathcad will assume an increment of 1. To illustrate, type "x:6;9" followed by "x=". The result is,

$x := 6..9$ $x =$

6
7
8
9

1.3 Graphics

One of the most useful things about a computational software package such as Mathcad is the ability to easily create graphs of functions. As we will see, these graphs allow one to gain a lot of insight into a problem by observing how a solution changes as some parameter (the magnitude of a load, an angle, a dimension etc.) is varied. This is so important that practically every problem in this supplement will contain at least one plot. By the time you have finished reading this supplement you should be very proficient at plotting in Mathcad. This section will introduce you to the basics of plotting in Mathcad.

Mathcad has the capability of creating a number of different types of graphs. Here we will consider only the X-Y plot. The most common and easiest way to generate a plot of a function is to use range variables. The following example will guide you through the basic procedure.

First define the function to be plotted. Type "f(x):x*exp(-x^2)"

$$f(x) := x \cdot exp\left(-x^2\right)$$

Now define a range variable covering the range over which you would like to plot the function.

$$x := -3, -2.9 .. 3$$

Now click at the desired location on the worksheet and insert an X-Y plot by (a) selecting *Insert...Graph...X-YPlot* from the main menu, (b) using the shortcut key "@" or (c) selecting the X-Y Plot icon from the graph toolbar. You should see an empty graph like the following.

You should see two empty placeholders on the x and y axes. By default, the insertion point should already be on the x placeholder. If not, click on that placeholder and type "x". Now click on the y placeholder and type "f(x)". After clicking away you should see the following graph.

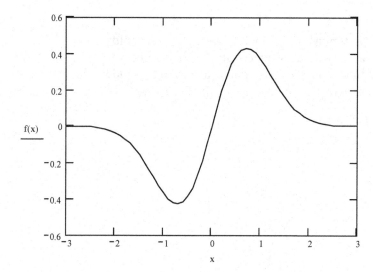

Parametric Studies

One of the most important uses of the computer in studying Mechanics is the convenience and relative simplicity of conducting parametric studies (not to be confused with parametric plotting discussed below). A parametric study seeks to understand the effect of one or more variables (parameters) upon a general solution. This is in contrast to a typical homework problem where you generally want to find one solution to a problem under some specified conditions. For example, in a typical homework problem you might be asked to find the reactions at the supports of a structure with a concentrated force of magnitude 200 lb that is oriented at an angle of 30 degrees from the horizontal. In a parametric study of the same problem you might typically find the reactions as a function of two parameters, the magnitude of the force and its orientation. You might then be asked to plot the reactions as a function of the magnitude of the force for several different orientations. A plot of this type is very beneficial in visualizing the general solution to a problem over a broad range of variables as opposed to a single case.

Parametric studies generally require making multiple plots of the same function with different values of a particular parameter in the function. Following is a very simple example.

$$f(a,x) := 5 + x - 5 \cdot x^2 + a \cdot x^3$$

What we would like to do is gain some understanding of how f varies with both x and a. We will illustrate this by plotting f as a function of x for a = -1, 0, and 1. As before, we first define a range variable.

$$x := -5, -4.9 .. 5$$

Now bring up an empty X-Y plot by typing "@". Type "x" into the placeholder on the x axis and then click on the y axis placeholder. Now type "f(-1,x),f(0,x),f(1,x)". Note that each time you type a comma, a new placeholder appears. When you click away you should see something like the following.

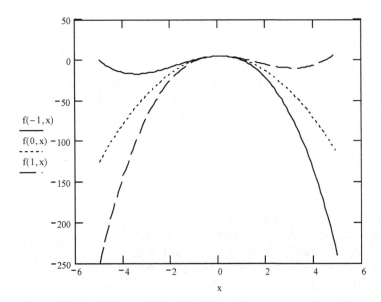

Parametric Plots

It often happens that one needs to plot some function y versus x but y is not known explicitly as a function of x. For example, suppose you know the x and y coordinates of a particle as a function of time but want to plot the trajectory of the particle, i.e. you want to plot the y coordinate of the particle versus the x coordinate. A plot of this type is generally called a parametric plot. Parametric plots are easy to obtain in Mathcad. You start by defining the two functions in terms of the common parameter and then define the common parameter as a range variable. Next, open an empty X-Y plot and type the two functions into the x and y axis placeholders. The following example illustrates this procedure.

$$f(a) := 10 \cdot a \cdot (2 - a)$$

In this example the parameter is a.

$$g(a) := \sin(3 \cdot a)$$

$$a := -1, -.95 .. 3.5$$

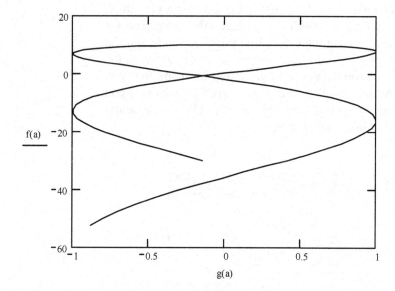

The range selected for the parameter can have a big, and sometimes surprising effect on the resulting graph. To illustrate, try increasing the upper limit on the range on a a few times and see how the graph changes.

You can, of course, also plot g as a function of f.

$a := -1, -.95 .. 6$

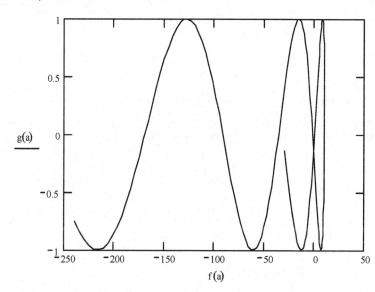

In the above examples we have more or less just accepted whatever graph Mathcad produced. This is the easiest approach and is certainly acceptable for many situations. You should be aware, though, that it is possible to change the appearance of a graph in several ways. To do this, first prepare your graph in the usual manner and then double click on it. You will get a pop-up menu that you can use to reformat the graph. At the top of the menu there are four tabs that you can select to alter different aspects of the graph's appearance. The figure below shows a menu where the "Traces" tab has been selected.

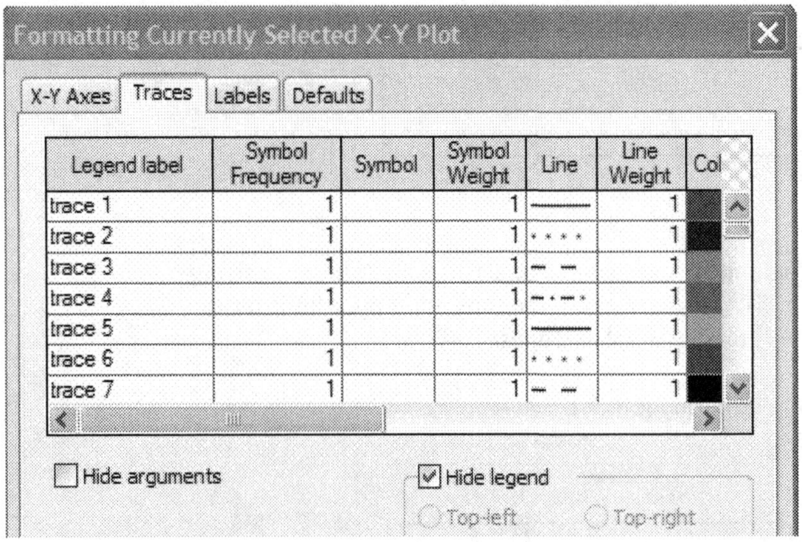

Here you can modify the line style, color and thickness (weight) of each curve. You can also plot symbols instead of lines. You should spend some time familiarizing yourself with the various graph formatting possibilities available in Mathcad.

1.4 Symbolic Math

Up to this point we have been using Mathcad essentially as a calculator. Well, obviously, a very sophisticated calculator, but a calculator nevertheless. There are times where it is very useful to have Mathcad perform mathematical calculations with symbols rather than numbers. This is very much like what you might do when deriving or manipulating equations in a homework problem. Except, Mathcad is less prone to making algebra mistakes.

Three of the most important applications of symbolic math will be discussed in the next three sections, namely symbolic vector algebra, symbolic calculus (integration and differentiation) and symbolic solution of one or more equations. The purpose of the present section is to introduce you to the basic procedures of symbolic math as well as to give a few other useful applications.

There are several approaches that can be used to perform symbolic mathematics. Here we will use just one primary approach and a slight modification of that approach. Start by opening the symbolic toolbar. This can be done either by selecting *View...Toolbars...Symbolic* from the main menu or by clicking the symbolic icon on the math toolbar. It looks like a graduation cap. Here's what you should see (note that the appearance might be slightly different in different versions of Mathcad).

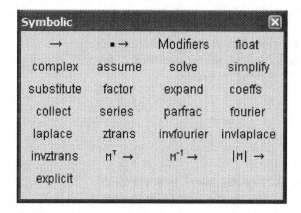

Let's start by illustrating symbolic simplification. First enter the expression you wish to simplify on the worksheet (no equals signs). Now click anywhere on this expression and then click on the *simplify* tab on the Symbolic tool bar. Finally, click anywhere on the worksheet and the simplified expression will appear. Here's a simple example.

$$\frac{\left(a \cdot x + b \cdot x^2\right)^2}{x^4}$$

Here's the expression we want to simplify. If you need help, type "(a*x+b*x^2)^2[space bar]/x^4". Now click anywhere on the expression and then press the *simplify* tab. After clicking away you should see the following.

$$\frac{\left(a \cdot x + b \cdot x^2\right)^2}{x^4} \text{ simplify } \rightarrow \frac{1}{x^2} \cdot (a + b \cdot x)^2$$

You can also simplify expressions containing previously defined functions. Here's another way to obtain the simplification above. See if you can reproduce it on your worksheet.

$$f(a, b, x) := a \cdot x + b \cdot x^2 \qquad g(x) := x^4$$

$$\frac{f(a, b, x)^2}{g(x)} \text{ simplify } \rightarrow \frac{1}{x^2} \cdot (a + b \cdot x)^2$$

Another useful symbolic operation is substitution. The substitution operator allows you to substitute an expression for a variable in another expression. Start with the expression you would like to substitute into. Click anywhere on this expression and then click the "substitute" tab on the Symbolic tool bar. You will get a bold equal sign with placeholders on either side. Fill in the placeholders so that you have *variable1=variable2*, where *variable2* is to be substituted for *variable1*. The following example illustrates the substitute operator.

$$\frac{\left(a \cdot x + 4 \cdot x^2\right)^2}{x^2 + a}$$

Start with the expression into which you would like to substitute. Click anywhere on this expression and then click the "substitute" tab on the Symbolic toolbar. You should see something like the following.

$$\frac{\left(a \cdot x + 4 \cdot x^2\right)^2}{x^2 + a} \text{ substitute, } \blacksquare = \blacksquare \rightarrow$$

Click in the left placeholder and type "a". Now click in the right placeholder and type "1+x^2". You should see the following result.

$$\frac{(a \cdot x + 4 \cdot x^2)^2}{x^2 + a} \quad \text{substitute}, a = 1 + x^2 \rightarrow \frac{\left[(1 + x^2) \cdot x + 4 \cdot x^2\right]^2}{(2 \cdot x^2 + 1)}$$

It is also possible to substitute previously defined functions.

$$f(x) := x^2 + 1$$

$$\frac{(2 \cdot x + 4 \cdot x^2)^2}{x^2 + 2} \quad \text{substitute}, x = f(x) \rightarrow \frac{\left[2 + 2 \cdot x^2 + 4 \cdot (1 + x^2)^2\right]^2}{\left[(1 + x^2)^2 + 2\right]}$$

The results following a substitution are often rather messy. To simplify, one could always copy the final result into the clipboard, paste it onto the worksheet, and then follow the procedure above to simplify. It is also possible to do several symbolic operations at once. The following example shows a substitution followed by a simplification. The procedure is the same as for substitution with one difference. After filling out the before and after placeholders, click on the "simplify" tab **before** clicking away.

$$\frac{(2 \cdot x + 4 \cdot x^2)^2}{x^2 + 2} \quad \begin{vmatrix} \text{substitute}, x = f(x) \\ \text{simplify} \end{vmatrix} \rightarrow 4 \cdot \frac{(3 + 5 \cdot x^2 + 2 \cdot x^4)^2}{(3 + 2 \cdot x^2 + x^4)}$$

Finally, here is an example with two substitutions followed by simplification.

$$\frac{(a \cdot x + b \cdot x^2)^2}{x^5} \quad \begin{vmatrix} \text{substitute}, a = x^3 + \tan(x) \\ \text{substitute}, b = -x^2 \\ \text{simplify} \end{vmatrix} \rightarrow \frac{1}{x^3} \cdot \tan(x)^2$$

1.5 Vector Algebra

The main application of vector algebra is in three dimensional problems where the geometry is difficult to visualize. Some of these difficulties include finding the x, y, and z components of a vector, moment arms for a force, projections of a force onto a line etc. The most useful vector operations are finding the magnitude of a vector and finding the dot or cross product of two vectors.

First we need to learn how to represent a vector in Mathcad. Start by opening the *Vector and Matrix Toolbar*. You can do this by selecting *View...Toolbars...Matrix* or by clicking the matrix icon on the *Math Toolbar* (it looks like a 3x3 matrix). Cartesian vectors are represented by three element column matrices. The following example shows how to create a vector.

Start by typing "u:". Now click on the matrix icon on the *Vector and Matrix Toolbar*. You can also select *Insert...Matrix* or use the shortcut **CNTRL+M**. In the popup menu select 3 rows and 1 column. After clicking OK, you should see the following

$$u := \begin{pmatrix} \blacksquare \\ \blacksquare \\ \blacksquare \end{pmatrix}$$

Now fill in the placeholders with the x, y, and z components of the vector. For example,

$$u := \begin{pmatrix} 3 \\ -2 \\ 6 \end{pmatrix}$$

Once the vector has been defined you can refer to the components of the vectors by typing the name of the vector with an index. Indices start at 0 in Mathcad so an index of 0, 1, and 2 correspond to x, y, and z. Indices are entered by typing "[". Don't confuse an index with a subscript, which is obtained by typing ".". For example, to print the y component of **u** type "u[1=".

$$u_1 = -2$$

To find the magnitude of **u**, select the absolute value icon ($|x|$) from the *Vector and Matrix Toolbar*. In the placeholder type "u=". You should see the following.

$$|u| = 7$$

In Statics, we often need to find unit vectors. A unit vector in the direction of **u** can be obtained by dividing **u** by the magnitude of **u**.

$$n := \frac{u}{|u|} \qquad\qquad n = \begin{pmatrix} 0.429 \\ -0.286 \\ 0.857 \end{pmatrix}$$

As an example of the above, suppose that you have a force **F** with magnitude 100 lb and with a line of action passing from point A (2, 0, 3) toward B (7, -2, 5). We can represent F as a Cartesian vector in Mathcad as follows.

$$r_A := \begin{pmatrix} 2 \\ 0 \\ 3 \end{pmatrix} \qquad\qquad r_B := \begin{pmatrix} 7 \\ -2 \\ 5 \end{pmatrix}$$

$$r_{AB} := r_A - r_B$$

$$F := 100 \cdot \frac{r_{AB}}{|r_{AB}|} \qquad\qquad F = \begin{pmatrix} -87.039 \\ 34.816 \\ -34.816 \end{pmatrix}$$

Dot and cross product operators can also be selected from the *Vector and Matrix Toolbar*. Shortcuts are ***** for dot product and **CNTRL+8** for cross product. Here are a few examples using the vectors we have already defined above.

$$u \cdot F = -539.641$$

$$r_A \times r_B = \begin{pmatrix} 6 \\ 11 \\ -4 \end{pmatrix}$$

$$M := r_A \times F \qquad\qquad M = \begin{pmatrix} -104.447 \\ -191.485 \\ 69.631 \end{pmatrix}$$

Vector operations can also be carried out symbolically. You will, of course, use the symbolic equals sign → instead of the evaluation equal sign =. Here are a few examples.

$$u := \begin{pmatrix} a \\ b \\ c \end{pmatrix} \qquad v := \begin{pmatrix} p \\ 3 \\ 4 \end{pmatrix} \qquad w := \begin{pmatrix} x \\ -5 \\ 4 \end{pmatrix}$$

After typing in the above three vectors you probably noticed that some of the variables appear in red since they have not been defined. This would obviously create a problem if you were going to evaluate some numerical results, however, it has no effect on symbolic calculations as can be seen from the following.

$$u \cdot v \rightarrow a \cdot p + 3 \cdot b + 4 \cdot c$$

$$\frac{v}{|v|} \rightarrow \begin{bmatrix} \dfrac{p}{\left[(|p|)^2 + 25\right]^{\left(\frac{1}{2}\right)}} \\[20pt] \dfrac{3}{\left[(|p|)^2 + 25\right]^{\left(\frac{1}{2}\right)}} \\[20pt] \dfrac{4}{\left[(|p|)^2 + 25\right]^{\left(\frac{1}{2}\right)}} \end{bmatrix}$$

$$u \times v \rightarrow \begin{pmatrix} 4 \cdot b - 3 \cdot c \\ c \cdot p - 4 \cdot a \\ 3 \cdot a - b \cdot p \end{pmatrix}$$

$$w \cdot (u \times v) = 0 \text{ solve}, x \rightarrow \frac{-(-5 \cdot c \cdot p + 32 \cdot a - 4 \cdot b \cdot p)}{(4 \cdot b - 3 \cdot c)}$$

1.6 Differentiation and Integration

Mechanics problems often require integration and/or differentiation. In Mathcad, you can perform these operations either numerically or symbolically. Before we get started you will want to open the Calculus Toolbar. You can open this by pressing the icon in the math toolbar or by selecting *View...Toolbars* from the main menu. The icons we will be using are those for the first and nth derivative and the definite and indefinite integral. The definite integral has *a* and *b* as integration limits. You may also want to open the Symbolic Toolbar.

Let's get started with a simple example of symbolic differentiation. Start by selecting the icon for the first derivative. Here's what you should see.

$$\frac{d}{d\blacksquare}\blacksquare$$

Note that there are two placeholders. Into the placeholder on the right hand side type the expression that you would like to differentiate (for this example, type "(a*sec(b*t))"). Then click on the placeholder in the denominator and enter the variable that you would like to differentiate with respect to. You should see the following.

$$\frac{d}{dt}(a \cdot \sec(b \cdot t))$$

Now click anywhere on this expression and click on the symbolic evaluation icon ($\rightarrow$) in the Symbolic Toolbar. After clicking away you will see the result of the symbolic differentiation.

$$\frac{d}{dt}(a \cdot \sec(b \cdot t)) \rightarrow a \cdot \sec(b \cdot t) \cdot \tan(b \cdot t) \cdot b$$

Higher order derivatives follow the same procedure except that there is an additional placeholder to fill in for the order of differentiation. See if you can reproduce the following result.

$$\frac{d^3}{dx^3} a \cdot \ln(b + x) \rightarrow 2 \cdot \frac{a}{(b + x)^3}$$

You can also use derivatives in defining functions. As an example, suppose a particle moves in a straight line and its position s is known as a function of time. From your elementary physics course you probably know that the first and second derivatives of the position give the velocity and acceleration of the particle.

$s(t) := 10 \cdot t - 20 \cdot t^2 + 2 \cdot t^3$

$v(t) := \dfrac{d}{dt} s(t)$

$a(t) := \dfrac{d^2}{dt^2} s(t)$

> Note the *assignment* equals sign ":=".

Now you can evaluate the velocity and acceleration at any time.

$v(1) = -24$ $a(1) = -28$

$v(7) = 24$ $a(7) = 44$

> Note the *evaluation* equals sign "=".

You can also plot the results.

$t := 0, 0.1 .. 10$

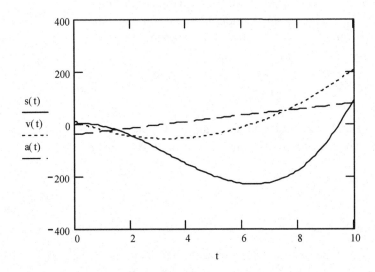

While the above is very convenient, especially when you want to numerically evaluate or plot the results after differentiation, it fails to provide the symbolic results. If you would like to have a record of these you can consider something like the following.

$s(t) := 10 \cdot t - 20 \cdot t^2 + 2 \cdot t^3$ position

$v(t) := \dfrac{d}{dt} s(t)$ velocity, $\dfrac{d}{dt} s(t) \rightarrow 10 - 40 \cdot t + 6 \cdot t^2$

$a(t) := \dfrac{d^2}{dt^2} s(t)$ acceleration, $\dfrac{d^2}{dt^2} s(t) \rightarrow -40 + 12 \cdot t$

or, $\dfrac{d}{dt} v(t) \rightarrow -40 + 12 \cdot t$

The procedure for performing integrations is very similar to that for differentiation. For example, to perform a symbolic integration: (a) click on the icon for either a definite or indefinite integral (or use the shortcut key **Shift+7** (or **&**) for definite and **Cntrl+i** for indefinite), (b) fill in the placeholders, (c) click anywhere on the expression and (d) click the Symbolic Evaluation icon (or use the short cut **Cntrl+.**). See if you can reproduce the following integrals.

$\displaystyle \int \sin(b \cdot x)\, dx \rightarrow \frac{-\cos(b \cdot x)}{b}$ $\displaystyle \int_c^d \sin(b \cdot x)\, dx \rightarrow \frac{-\cos(d \cdot b)}{b} + \frac{\cos(c \cdot b)}{b}$

$\displaystyle \int \ln(x)\, dx \rightarrow x \cdot \ln(x) - x$ $\displaystyle \int_c^d \ln(x)\, dx \rightarrow d \cdot \ln(d) - d - c \cdot \ln(c) + c$

If a definite integral contains no unknown parameters either in the integrand or the integration limits, the above procedure will provide numerical answers. Here are a few examples.

$\displaystyle \int_0^3 \left(x + 3 \cdot x^3 \right) dx \rightarrow \frac{261}{4}$ $\displaystyle \int_2^5 \ln(x)\, dx \rightarrow 5 \cdot \ln(5) - 3 - 2 \cdot \ln(2)$

Note that Mathcad will try to return an exact result when the Symbolic Evaluation procedure is used. This results in fractions or functions as in the above examples. This is very useful in some situations, however, one often wants to know the numerical answer without having to evaluate a result such as the above with a calculator. Thus, Mathcad also allows you to obtain results for numerical integration as floating point numbers. This can be accomplished by following the same procedure outlined above except that for step (d) you press the equals sign "=" on your keyboard instead of clicking the Symbolic Evaluation icon. To illustrate, we will repeat the same two integrals above.

$$\int_0^3 \left(x + 3 \cdot x^3\right) dx = 65.25 \qquad\qquad \int_2^5 \ln(x) \, dx = 3.661$$

1.7 Solving Equations

Solving a single equation symbolically can be accomplished in a manner very similar to other symbolic operations considered earlier. As an example, try typing the following equation on to your worksheet, being sure to type **Cntrl** = for the equals sign (you should see a bold equals sign =).

$$a \cdot x^2 + b \cdot x + c = 0$$

Click anywhere on the equation and then click *solve* on the *Symbolic Toolbar* . Type the variable you wish to solve for (in this example x) in the placeholder and the click away. You should see the following.

$$a \cdot x^2 + b \cdot x + c = 0 \text{ solve}, x \rightarrow \begin{bmatrix} \dfrac{1}{(2 \cdot a)} \cdot \left[-b + \left(b^2 - 4 \cdot a \cdot c\right)^{\left(\frac{1}{2}\right)} \right] \\ \dfrac{1}{(2 \cdot a)} \cdot \left[-b - \left(b^2 - 4 \cdot a \cdot c\right)^{\left(\frac{1}{2}\right)} \right] \end{bmatrix}$$

Note that Mathcad has found both solutions to the (hopefully) familiar quadratic equation. If the equals sign is omitted, Mathcad will assume that the expression is set equal to zero, i.e. Mathcad will find the roots of the expression. Here is an alternative way to obtain the above result.

$$a \cdot x^2 + b \cdot x + c \text{ solve}, x \rightarrow \begin{bmatrix} \dfrac{1}{(2 \cdot a)} \cdot \left[-b + \left(b^2 - 4 \cdot a \cdot c\right)^{\left(\frac{1}{2}\right)} \right] \\ \dfrac{1}{(2 \cdot a)} \cdot \left[-b - \left(b^2 - 4 \cdot a \cdot c\right)^{\left(\frac{1}{2}\right)} \right] \end{bmatrix}$$

If the variable being solved for is the only unknown in the equation, Mathcad will return a number as the result. Here are a couple of examples.

$$2 \cdot x^2 + 4 \cdot x - 12 \text{ solve}, x \rightarrow \begin{pmatrix} -1 + \sqrt{7} \\ -1 - \sqrt{7} \end{pmatrix}$$

$$5 \cdot \sin(\theta) - \cos(\theta) \equiv 1 \text{ solve}, \theta \rightarrow \begin{pmatrix} \pi \\ \text{atan}\left(\dfrac{5}{12}\right) \end{pmatrix}$$

You can also solve equations using **_Given...Find_**. For the symbolic case, one starts with the basic _Given...Find_ format shown below.

Given

$$a \cdot x^2 + b \cdot x + c \equiv 0$$

Find(x)

Now click on "Find(x)" and then click on the _Symbolic Evaluation_ icon ($\rightarrow$). After clicking away you should see the following result.

Given

$$a \cdot x^2 + b \cdot x + c \equiv 0$$

$$\text{Find}(x) \rightarrow \left[\frac{1}{(2 \cdot a)} \cdot \left[-b + \left(b^2 - 4 \cdot a \cdot c\right)^{\left(\frac{1}{2}\right)} \right] \quad \frac{1}{(2 \cdot a)} \cdot \left[-b - \left(b^2 - 4 \cdot a \cdot c\right)^{\left(\frac{1}{2}\right)} \right] \right]$$

For a numerical solution you would use the same procedure but type "Find(x)=". Here's an example.

$$g(x) := 2 \cdot x^2 + 1 - 10 \cdot \sin(x)$$

$$x := 0$$

Given

$$g(x) \equiv 0$$

$$\text{Find}(x) = 0.102$$

$x := 2$

Given

$g(x) = 0$

$\text{Find}(x) = 2.008$

Given...Find can also be used to solve simultaneous equations either symbolically or numerically. The approach is essentially the same as that described above for a single equation except, of course, that more than one equation will appear between the *Given* and *Find* statements. Also, for numerical solutions, an initial guess should be provided for all unknowns. Following are several examples. An easy way to tell at a glance whether the solution is symbolic or numerical is to see whether the symbolic evaluation symbol ($\rightarrow$) appears after *Find*.

Given

$-P \cdot \sin(\beta) + B_x - A_x = 0$

$A_y + P \cdot \cos(\beta) - w \cdot a = 0$

> A subscript can be obtained by typing "." before the subscript. For example, by typing "A.x".

$P \cdot a \cdot \cos(\beta) - P \cdot b \cdot \sin(\beta) + B_x \cdot c - \dfrac{1}{2} \cdot w \cdot a^2 = 0$

$$\text{Find}(A_x, A_y, B_x) \rightarrow \begin{bmatrix} \dfrac{1}{2} \cdot \dfrac{\left(-2 \cdot P \cdot \sin(\beta) \cdot c - 2 \cdot P \cdot a \cdot \cos(\beta) + 2 \cdot P \cdot b \cdot \sin(\beta) + w \cdot a^2\right)}{c} \\ -P \cdot \cos(\beta) + w \cdot a \\ \dfrac{1}{2} \cdot \dfrac{\left(-2 \cdot P \cdot a \cdot \cos(\beta) + 2 \cdot P \cdot b \cdot \sin(\beta) + w \cdot a^2\right)}{c} \end{bmatrix}$$

Given

$x^2 + y^2 = 12 \qquad x \cdot y = 4$

$$\text{Find}(x,y) \rightarrow \begin{pmatrix} \sqrt{5}-1 & -1-\sqrt{5} & \sqrt{5}+1 & 1-\sqrt{5} \\ \sqrt{5}+1 & 1-\sqrt{5} & \sqrt{5}-1 & -1-\sqrt{5} \end{pmatrix}$$

Note that each column in the last result represents a solution. Thus, in the last example, Mathcad has found four solutions, the first being $x = \sqrt{5}-1$ and $y = \sqrt{5}+1$.

$x := 0 \quad y := 0 \quad z := 0$

> This is our initial guess for a numerical solution

Given

$x^2 + y = 12 \qquad x \cdot y = 4 \qquad x - y = z$

$$\text{Find}(x,y,z) = \begin{pmatrix} 3.284 \\ 1.218 \\ 2.065 \end{pmatrix}$$

$x := 0 \quad y := 5 \quad z := 5$

> Now let's try another guess for the same set of equations.

Given

$x^2 + y = 12 \qquad x \cdot y = 4 \qquad x - y = z$

$$\text{Find}(x,y,z) = \begin{pmatrix} 0.337 \\ 11.887 \\ -11.55 \end{pmatrix}$$

*Finding Numerical Solutions with **root***

Numerical solutions to single equations can be obtained with root. This is particularly useful in those situations where solve fails to find a solution. We will illustrate by finding a numerical solution to the equation 10sin(x) = 2x^2 + 1.

Before using the **root** function you should first provide a value in the neighborhood of the solution you are seeking. This is especially important if, as in the present case, there is more than one solution. Well, you may be wondering how we can determine the neighborhood of a solution if we do not yet know the solution. Actually, this is very easy to do. First we define a function g(x) whose roots will be the solution to the equation of interest. Next, we plot this function in order to estimate the location of points where g(x) = 0.

$$g(x) := 2 \cdot x^2 + 1 - 10 \cdot \sin(x)$$

$$x := -1, -0.9 .. 3$$

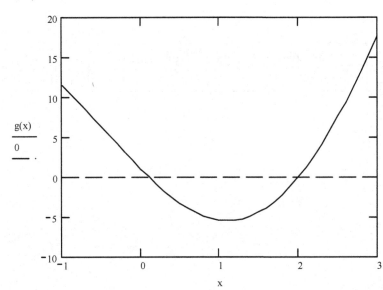

From the graph above we see that $g(x) = 0$ in two places, about $x = 0$ and $x = 2$. These results provide our initial guesses for the root command. Here's how it works.

$x := 0 \quad \text{root}(g(x), x) = 0.102$

$x := 2 \quad \text{root}(g(x), x) = 2.008$

Or, equivalently,

$x := 0 \quad x1 := \text{root}(g(x), x) \qquad x1 = 0.102$

$x := 2 \quad x2 := \text{root}(g(x), x) \qquad x2 = 2.008$

FORCE SYSTEMS

2

This chapter introduces the basic properties of forces and moments in two and three dimensions. Problem 2.1 is a 2D problem that investigates the effects of the orientation of a force upon the resultant of two forces. The problem illustrates how the automatic substitution performed by a computer can simplify what might normally be a rather tedious algebra problem, a point also illustrated (even more dramatically) in problem 2.6. Problem 2.2 and 2.3 are parametric studies involving 2D moment and equivalent force-couple systems respectively. Problem 2.2 calculates the magnitude of a force using numerical substitution but also illustrates how to obtain the symbolic result. Problems 2.4 and 2.5 are 3D problems. Problem 2.4 uses symbolic vector algebra to obtain the projection of a force onto a line using the dot product. Problem 2.5 shows how to carry out a numerical and symbolic cross product with Mathcad and also involves an interesting design application. Problem 2.6 is another 2D problem and illustrates how to find the maximum value of a moment by setting the slope equal to zero. The slope is determined using symbolic differentiation and the location where the slope is zero is obtained with a symbolic *solve* operation. This problem also provides a good illustration of the power of symbolic substitution.

2.1 Problem 2/20 (2D Rectangular Components)

It is desired to remove the spike from the timber by applying force along its horizontal axis. An obstruction A prevents direct access, so that two forces, one 400 lb and the other **P** are applied by cables as shown. Here we want to investigate the effects of the distance between the spike and the obstruction on the two forces so replace 8" by d in the figure. Determine the magnitude of **P** necessary to insure a resultant **T** directed along the spike. Also find T.   Plot P and T as a function of d letting d range between 2 and 12 inches.

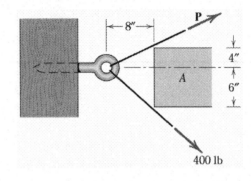

Problem Formulation

The two forces and their resultant are shown on the diagram to the right. The horizontal component of the resultant is T while the vertical component is 0. Thus,

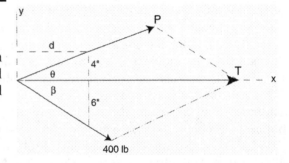

$$T = \Sigma F_x = P\cos\theta + 400\cos\beta$$

$$0 = \Sigma F_y = P\sin\theta - 400\sin\beta$$

These two equations can be easily solved for P and T.

$$P = \frac{400\sin\beta}{\sin\theta} \qquad T = 400(\sin\beta\cot\theta + \cos\beta)$$

At this point we have P and T as functions of β and θ. From the figure above we can relate β and θ to d as follows.

$$\theta = \tan^{-1}(4/d) \qquad \beta = \tan^{-1}(6/d)$$

One nice thing about using a computer is that it will not be necessary to substitute these results into those above to get P and T explicitly as functions of d. The computer carries out this substitution automatically.

Mathcad Worksheet

$$\theta(d) := \operatorname{atan}\left(\frac{4}{d}\right) \qquad \beta(d) := \operatorname{atan}\left(\frac{6}{d}\right)$$

$$P(d) := 400 \cdot \frac{\sin(\beta(d))}{\sin(\theta(d))}$$

$$T(d) := 400 \cdot (\sin(\beta(d)) \cdot \cot(\theta(d)) + \cos(\beta(d)))$$

$$d := 2, 2.05 .. 12$$

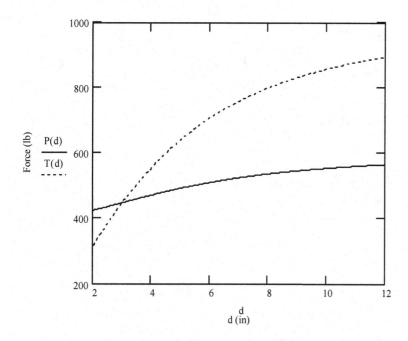

2.2 Problem 2/53 (2D Moment)

The masthead fitting supports the two forces shown. Here we want to investigate the effects of the orientation of the 5 kN load so replace the angle 30° by θ. Find the magnitude of **T** which will cause no bending of the mast (zero moment) at O. For this T, determine the magnitude of the resultant **R** of the two forces. Plot T and R as a function of θ letting q vary between 0 and 180°.

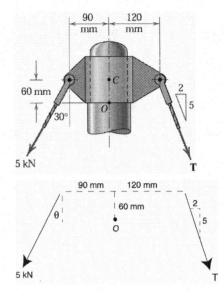

Problem Formulation

Setting the moment about O to zero yields (note that $\sqrt{2^2 + 5^2} = \sqrt{29}$)

$$M_0 = 5\left[\cos\theta(90) + \sin\theta(60)\right] - T\left[\frac{5}{\sqrt{29}}(120) + \frac{2}{\sqrt{29}}(60)\right] = 0$$

Solving, $T = \dfrac{5\left[\cos\theta(90) + \sin\theta(60)\right]}{\dfrac{5}{\sqrt{29}}(120) + \dfrac{2}{\sqrt{29}}(60)} = \dfrac{5\sqrt{29}}{24}\left(3\cos\theta + 2\sin\theta\right)$

Now we can find the resultant R in terms of T,

$$R_x = \frac{2}{\sqrt{29}}T - 5\sin\theta \ (\rightarrow) \quad R_y = \frac{5}{\sqrt{29}}T + 5\cos\theta \ (\downarrow) \qquad R = \sqrt{R_x^2 + R_y^2}$$

We'll let Mathcad substitute for T and plot T and R as a function of θ.

Mathcad Worksheet

$$T(\theta) := \frac{5 \cdot \sqrt{29}}{24} \cdot (3 \cdot \cos(\theta) + 2 \cdot \sin(\theta))$$

$$R_x(\theta) := \frac{2}{\sqrt{29}} \cdot T(\theta) - 5 \cdot \sin(\theta) \qquad R_y(\theta) := \frac{5}{\sqrt{29}} \cdot T(\theta) + 5 \cdot \cos(\theta)$$

$$R(\theta) := \sqrt{R_x(\theta)^2 + R_y(\theta)^2}$$

There may be situations where you might like to know explicitly how R depends upon theta. If this is the case you can use symbolic algebra to evaluate R as illustrated below.

$$\sqrt{R_x(\theta)^2 + R_y(\theta)^2} \rightarrow \left[\left(\frac{5}{4} \cdot \cos(\theta) - \frac{25}{6} \cdot \sin(\theta) \right)^2 + \left(\frac{65}{8} \cdot \cos(\theta) + \frac{25}{12} \cdot \sin(\theta) \right)^2 \right]^{\left(\frac{1}{2}\right)}$$

Of course, this result is not necessary to make a plot of R versus theta. Mathcad will automatically make the substitutions numerically in order to provide the graphical results. Note the conversion to degrees in the graph below.

$$\theta := 0, 0.01 .. \pi$$

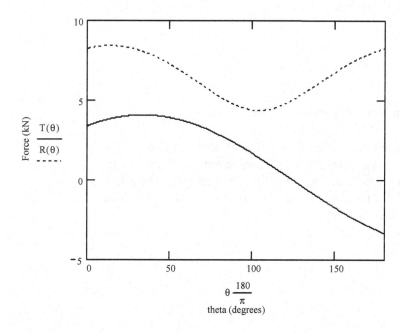

2.3 Problem 2/88 (2D Resultants)

The directions of the two thrust vectors of an experimental aircraft can be independently changed form the conventional forward direction within limits. Here we want to investigate the effects of the orientation of one of the thrusts so change the 15° angle in the figure to θ. For this case determine the equivalent force-couple system at O. Then replace this force-couple system by a single force and specify the point on the x-axis through which the line of action of this resultant passes. Plot (a) the ratio of the magnitude of the resultant to that of the thrust (R/T) and (b) the distance x where the resultant intersects the x-axis as functions of θ for θ between 0 and 30°.

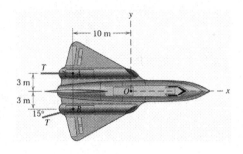

Problem Formulation

For an equivalent force-couple system at O we have,

$$R_x = T + T\cos\theta = T(1+\cos\theta) \quad R_y = T\sin\theta$$

$$R = \sqrt{R_x^2 + R_y^2} = T\sqrt{(1+\cos\theta)^2 + (\sin\theta)^2}$$

$$\frac{R}{T} = \sqrt{(1+\cos\theta)^2 + (\sin\theta)^2}$$

$$M_O = -T(3) + T\cos\theta(3) - T\sin\theta(3) = T(3\cos\theta - 3 - 10\sin\theta) \circlearrowleft$$

Now we want a single resultant force that is equivalent to the force-couple system above. The easiest way to visualize this is to imagine moving the resultant from O to the right along the x-axis. As you do this you generate a counter-clockwise moment about O of $R_y x$ (R_x does not produce a moment about O). The condition for determining x is thus $M_O = R_y x$.

$$x = \frac{M_O}{R_y} = \frac{(3\cos\theta - 3 - 10\sin\theta)}{\sin\theta}$$

Mathcad Worksheet

Let R_P be the ratio R/T

$$R_p(\theta) := \sqrt{(1 + \cos(\theta))^2 + \sin(\theta)^2}$$

$$\theta := 0, 0.01 .. \, 30 \cdot \frac{\pi}{180}$$

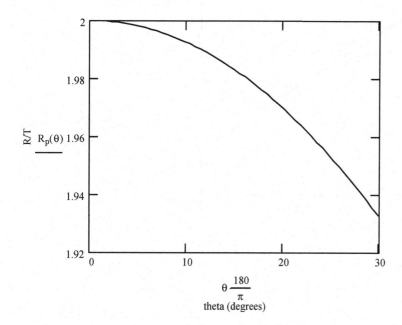

$$x(\theta) := \frac{3 \cdot \cos(\theta) - 3 - 10 \cdot \sin(\theta)}{\sin(\theta)}$$

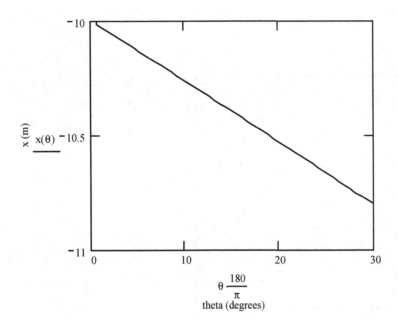

2.4 Problem 2/105 (3D Rect. Components)

The rigid pole and cross-arm assembly is supported by the three cables shown. A turnbuckle at D is tightened until it induces a tension T in CD of 1.2 kN. Let the distance OD be d instead of 3 m and determine the magnitude T_{GF} of the projection of **T** onto line GF. Plot T_{GF} as a function of d letting d vary between 0 and 20 meters.

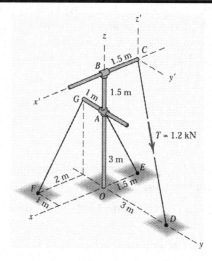

Problem Formulation

First we write **T** as a Cartesian vector and then find a unit vector along line GF.

$$\mathbf{T} = T\frac{1.5\mathbf{i} + d\mathbf{j} - 4.5\mathbf{k}}{\sqrt{(1.5)^2 + d^2 + (4.5)^2}} = (1.2)\frac{1.5\mathbf{i} + d\mathbf{j} - 4.5\mathbf{k}}{\sqrt{22.5 + d^2}}$$

$$\mathbf{n}_{GF} = \frac{2\mathbf{i} - 3\mathbf{k}}{\sqrt{(2)^2 + (3)^2}} = \frac{2\mathbf{i} - 3\mathbf{k}}{\sqrt{13}}$$

The projection T_{GF} is now found by taking the dot product of the above two vectors.

$$T_{GF} = \mathbf{T} \bullet \mathbf{n}_{GF} = \frac{1.2(1.5)(2) + 1.2d(0) + 1.2(-4.5)(-3)}{\sqrt{13}\sqrt{22.5 + d^2}} = \frac{19.8/\sqrt{13}}{\sqrt{22.5 + d^2}}$$

Although the vector operations are fairly simple in this problem, we will go ahead and let Mathcad do them for purposes of illustration. Considerably more involved vector operations can be found in problem 3.3 in the next chapter.

Mathcad Worksheet

$$r_{CD}(d) := \begin{pmatrix} 1.5 \\ d \\ -4.5 \end{pmatrix}$$

Above is the vector r_{CD} that will be used below to obtain the vector **T**. This is a three-element column vector and is obtained as follows. After typing "r.CD:" (the

"." reduces CD to a subscript) click *Insert...Matrix*. In the popup menu set the number of rows to 3 and the number of columns to 1. After clicking *insert* you will see the column vector with three placeholders. Fill the placeholders with the x, y, and z components of the vector. You can also insert a matrix from the *Matrix toolbar*.

$$T(d) := \frac{1.2 \cdot r_{CD}(d)}{\left| r_{CD}(d) \right|}$$

Use the *Matrix Toolbar* to get the absolute value operator | |, which computes the magnitude of the vector r_{CD}. You could also type in the magnitude of r_{CD} (= $\sqrt{22.5 + d^2}$) but letting Mathcad do this is a little more convenient.

$$n_{GF} := \frac{1}{\sqrt{13}} \cdot \begin{pmatrix} 2 \\ 0 \\ -3 \end{pmatrix}$$

$$T_{GF}(d) := T(d) \cdot n_{GF}$$

To take the dot product of the two vectors, use the usual multiplication symbol *. If you want the explicit relation between T_{GF} and d, use the symbolic operator $\rightarrow$

$$T(d) \cdot n_{GF} \rightarrow \frac{1.5230769230769230769}{\left[22.50 + \left(|d| \right)^2 \right]^{\left(\frac{1}{2} \right)}} \cdot \sqrt{13}$$

$d := 0, 0.05 .. 20$

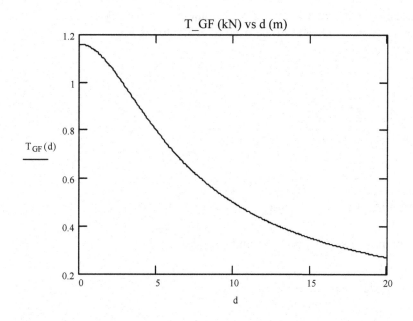

T_GF (kN) vs d (m)

2.5 Sample Problem 2/13 (3D Moment)

A tension **T** of magnitude 10 kN is applied to the cable attached to the top A of the rigid mast and secured to the ground at B. Let the coordinates of point B be $(x_B, 0, z_B)$ instead of $(12, 0, 9)$ as in the figure. Find a general expression for $\mathbf{M_O}$ (the moment of **T** about the base O) as a function of x_B and z_B. (a) Plot the magnitude of $\mathbf{M_O}$ (M_O) and its components about the x and z axes (M_x and M_z) as a function of x_B for $z_B = 9$ m. For this case, determine the suitable range for x_B if M_O is not to exceed 100 kN·m. (b) Find a simple way to explain to a construction worker (who has not taken this course) all safe locations for B if M_O cannot exceed 100 kN·m.

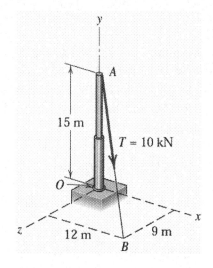

Problem Formulation

First, we express $\mathbf{T}$ as a Cartesian vector and then find the moment by taking a cross product. In the worksheet below, Mathcad will be used to evaluate the cross product.

$$\mathbf{T} = T\mathbf{n}_{AB} = 10\left[\frac{x_B\mathbf{i} - 15\mathbf{j} + z_B\mathbf{k}}{\sqrt{x_B^2 + 15^2 + z_B^2}}\right]$$

$$\mathbf{M_O} = \mathbf{r_{OA}} \times \mathbf{T} = 15\mathbf{j} \times 10\left[\frac{x_B\mathbf{i} - 15\mathbf{j} + z_B\mathbf{k}}{\sqrt{x_B^2 + 15^2 + z_B^2}}\right] = 150\left[\frac{z_B\mathbf{i} - x_B\mathbf{k}}{\sqrt{x_B^2 + 15^2 + z_B^2}}\right]$$

(a) M_x and M_z are the scalar components of $\mathbf{M_O}$ in the x and z-directions respectively.

$$M_x = \frac{150z_B}{\sqrt{x_B^2 + 15^2 + z_B^2}} \quad M_z = \frac{-150x_B}{\sqrt{x_B^2 + 15^2 + z_B^2}}$$

$$M_O = \sqrt{M_x^2 + M_z^2} = 150\sqrt{\frac{x_B^2 + z_B^2}{x_B^2 + 15^2 + z_B^2}}$$

After substituting $z_B = 9$, M_O, M_x, and M_z can be plotted as a function of x_B. The result can be found in the Mathcad worksheet that follows. To find the acceptable range of x_B given $M_O \leq 100$ kN•m, substitute $M_O = 100$ and $z_B = 9$ into the equation for M_O and then solve for x_B. The result is $x_B = \pm 3\sqrt{11}$ m. Thus, the acceptable range is

$$-3\sqrt{11} \leq x_B \leq 3\sqrt{11} \text{ m}$$

(b) Substituting $M_O = 100$ into the equation above gives,

$$100 = 150\sqrt{\frac{x_B^2 + z_B^2}{x_B^2 + 15^2 + z_B^2}} \; .$$

Squaring both sides of this equation and rearranging terms yields

$$x_B^2 + z_B^2 = 180$$

which is a circle of radius $\sqrt{180} = 13.42$ m (44 ft). Thus, the construction worker should be instructed to keep B within 44 feet of the mast.

Mathcad Worksheet

First we set up the vector expressions for **T** and $\mathbf{r}_{OA}$.

$$r_{AB}(xb, zb) := \begin{pmatrix} xb \\ -15 \\ zb \end{pmatrix} \qquad n_{AB}(xb, zb) := \frac{r_{AB}(xb, zb)}{\left| r_{AB}(xb, zb) \right|}$$

$$T(xb, zb) := 10 \cdot n_{AB}(xb, zb)$$

$$r_{OA} := \begin{pmatrix} 0 \\ 15 \\ 0 \end{pmatrix}$$

$$M_O(xb, zb) := r_{OA} \times T(xb, zb)$$

$$M_{OM}(xb, zb) := \left| M_O(xb, zb) \right|$$

Using indices 0 and 2 on $\mathbf{M}_O$ allows us to pick out the x and z components of the vector moment.

$$M_x(xb, zb) := M_O(xb, zb)_0$$

$$M_z(xb, zb) := M_O(xb, zb)_2$$

At this point we are ready to make a plot of M_{OM} (the magnitude of $\mathbf{M}_O$), M_x, and M_z as a function of xb with zb = 9. The problem statement asked for a general expression for **Mo** in terms of xb and zb. For this we need to evaluate the cross product using the symbolic operator.

$$r_{OA} \times T(xb, zb) \rightarrow \begin{bmatrix} 150 \cdot \dfrac{zb}{\left[(|xb|)^2 + 225 + (|zb|)^2 \right]^{\left(\frac{1}{2}\right)}} \\ 0 \\ -150 \cdot \dfrac{xb}{\left[(|xb|)^2 + 225 + (|zb|)^2 \right]^{\left(\frac{1}{2}\right)}} \end{bmatrix}$$

$xb := -15, -14.9 .. 15$

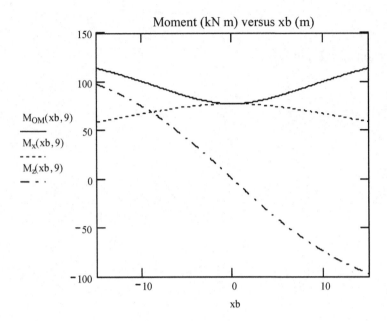

The following solves for the values of xb for which the magnitude of the moment will equal 100 kN m when zb = 9 m.

$$M_{OM}(x, 9) = 100 \text{ solve, } x \rightarrow \begin{bmatrix} 3 \cdot 11^{\frac{1}{2}} \\ (-3) \cdot 11^{\frac{1}{2}} \end{bmatrix}$$

Thus, the acceptable range is

$$-3\sqrt{11} \le x_B \le 3\sqrt{11} \text{ m}$$

2.6 Problem 2/181 (2D Moment)

A flagpole with attached light triangular frame is shown here for an arbitrary position during its raising. The 75-N tension in the erecting cable remains constant. Determine and plot the moment about the pivot O of the 75-N force for the range $0 \leq \theta \leq 90°$. Determine the maximum value of this moment and the elevation angle at which it occurs; comment on the physical significance of the latter. The effects of the diameter of the drum at D may be neglected.

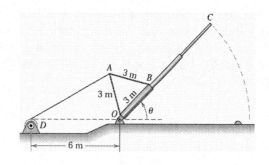

Problem Formulation

From the diagram to the right we can find the moment about O in two ways. The most obvious is to resolve T into horizontal and vertical components and then multiply by the respective moment arms. The result is,

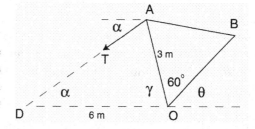

$$M_O = T\cos\alpha(3\sin\gamma) + T\sin\alpha(3\cos\gamma) = 3T(\cos\alpha\sin\gamma + \sin\alpha\cos\gamma)$$

A simpler expression can be found by first sliding T along its line of action to D. Once this has been done we see that only the vertical component ($T\sin\alpha$) produces a moment about O with moment arm of 6 m. Thus,

$$M_O = 6T\sin\alpha$$

Here we will use the second, simpler, expression. As an exercise you may want to verify that the two expressions yield identical results.

Now we have to do a little geometry to relate α to θ. From the diagram we have,

$$\gamma + 60^0 + \theta = 180^0 \qquad \gamma = 120^0 - \theta \qquad \gamma = \frac{2\pi}{3} - \theta \text{ (radians)}$$

$$AD = \sqrt{6^2 + 3^2 - 2(6)(3)\cos\gamma} \qquad\qquad \text{(law of cosines)}$$

$$\frac{\sin\alpha}{3} = \frac{\sin\gamma}{AD} \qquad \alpha = \sin^{-1}\left(\frac{3\sin\gamma}{AD}\right) \qquad \text{(law of sines)}$$

This problem illustrates very well one of the advantages of a computer solution. At this point you should be sure that you understand that we now have the moment as a function of θ. The reason is that we have M_O as a function of α, α as a function of γ and γ as a function of θ. It is not necessary to actually make the substitutions ourselves, the computer will do that for us. In addition to this, Mathcad, unlike many other programs, can also make the substitution symbolically. For example, from the worksheet below, we have the following symbolic result for M_O,

$$M_O = \frac{450 \sin\left(\frac{\pi}{3} + \theta\right)}{\sqrt{5 + 4\cos\left(\frac{\pi}{3} + \theta\right)}}$$

The maximum moment and the elevation at which it occurs are determined in the worksheet below. The angle θ is determined by solving the equation $dM_O / d\theta = 0$. Substitution of this result into M_O gives the maximum moment. The results are $M_{Omax} = 225$ N-m at $\theta = \pi/3 = 60°$.

It turns out that we could have anticipated this result with some simple geometry. First, we know that the maximum moment will occur when **T** is perpendicular to *OA*. This yields the right triangle shown to the right. From this diagram we see that $\cos\gamma = 3/6$ so that $\gamma = 60°$. From our results above, $\theta = 60°$ when $\gamma = 60°$. Also, for this orientation, $M_O = 3(75) = 225$ N-m.

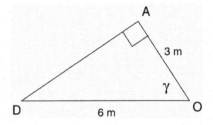

Mathcad Worksheet

$$\gamma(\theta) := \frac{2 \cdot \pi}{3} - \theta$$

$$AD(\theta) := \sqrt{6^2 + 3^2 - 2 \cdot 6 \cdot 3 \cdot \cos(\gamma(\theta))}$$

$$\alpha(\theta) := asin\left(\frac{3 \cdot \sin(\gamma(\theta))}{AD(\theta)}\right)$$

$$M_O(\theta) := 6 \cdot 75 \cdot \sin(\alpha(\theta))$$

To obtain Mo explicitly as a function of theta, carry out the last operation symbolically.

$$6 \cdot 75 \cdot \sin(\alpha(\theta)) \rightarrow 450 \cdot \frac{\sin\left(\frac{1}{3} \cdot \pi + \theta\right)}{\left(5 + 4 \cdot \cos\left(\frac{1}{3} \cdot \pi + \theta\right)\right)^{\left(\frac{1}{2}\right)}}$$

$$\theta := 0, 0.01 .. \frac{\pi}{2}$$

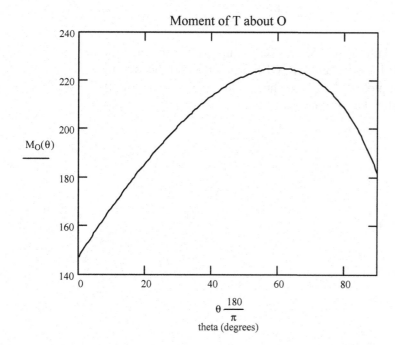

Moment of T about O

$\theta \dfrac{180}{\pi}$

theta (degrees)

To find the maximum moment we first differentiate Mo with respect to theta and set the result equal to zero. Solving this equation gives the angle theta at which the maximum occurs. This value is then substituted back into Mo to get the maximum.

$$dMO(\theta) := \frac{d}{d\theta} M_O(\theta)$$

$$dMO(x) = 0 \text{ solve}, x \rightarrow \begin{bmatrix} \dfrac{-1}{3} \cdot \pi + \text{atan2}\left(-2, i \cdot 3^{\frac{1}{2}}\right) \\ \dfrac{-1}{3} \cdot \pi + \text{atan2}\left[-2, (-i) \cdot 3^{\frac{1}{2}}\right] \\ \dfrac{1}{3} \cdot \pi \\ -\pi \end{bmatrix}$$

From the above we see that Mathcad has found four solutions to the equation. The first two are imaginary and can be ignored. Of the two real solutions, only one ($\pi/3 = 60$) is in the range plotted. We'll go ahead and evaluate Mo at both angles. The result indicates that the second real solution (-180 degrees) is a minimum.

$$M_O\left(\frac{\pi}{3}\right) = 225$$

$$M_O(-\pi) = -225$$

EQUILIBRIUM

3

This chapter considers equilibrium of two and three-dimensional structures and is the foundation for the study of engineering Statics. Problems 3.1 through 3.3 are two dimensional equilibrium problems while problem 3.4 involves equilibrium in three dimensions. Problem 3.1 is a fairly straightforward parametric study. Two equations are solved simultaneously for two unknowns using the symbolic *Given...Find* procedure. Problem 3.3 illustrates how to find the minimum value of a force by setting the slope equal to zero. The slope is determined using symbolic differentiation and the location where the slope is zero is obtained with a symbolic *solve* operation. Problem 3.4 involves some lengthy symbolic vector algebra that results in six scalar equilibrium equations. These equilibrium equations are solved symbolically using the *Given...Find* procedure.

3.1 Problem 3/26 (2D Equilibrium)

The indicated location of the center of gravity of the 3600-lb pickup truck is for the unladen condition. A load W_L whose center of gravity is x inches behind the rear axle is added to the truck. Find the relationship between W_L and x if the normal forces under the front and rear wheels are to be equal. For this case, plot W_L as a function of x for x ranging between 0 and 50 inches.

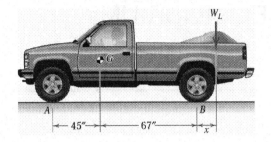

Problem Formulation

The free-body diagram for the truck is shown to the right. Normally, the two normal forces under the wheels would not be identical, of course. Here we want the relationship between the weight W_L and its location (x) which results in these two forces being equal. This relationship is found from the equilibrium equations.

$$\Sigma M_B = 0 = 3600(67) - N(112) - W_L x = 0$$

$$\Sigma F_y = 0 = N + N - 3600 - W_L = 0$$

The second equation can be solved for N and then substituted into the first equation to yield the required relation between W_L and x. This relation can then be solved for W_L.

Mathcad Worksheet

First we solve the two equilibrium equations for N and WL using the symbolic *Given...Find*.

Given

$$3600 \cdot 67 - N \cdot 112 - W_L \cdot x \equiv 0$$

$$2 \cdot N - 3600 - W_L \equiv 0$$

$$\text{Find}(N, W_L) \rightarrow \begin{bmatrix} 1800 \cdot \dfrac{(67 + x)}{(56 + x)} \\[2ex] \dfrac{39600}{(56 + x)} \end{bmatrix}$$

$$W_L(x) := \frac{39600}{(56 + x)}$$

$$x := 0, 0.1 .. 50$$

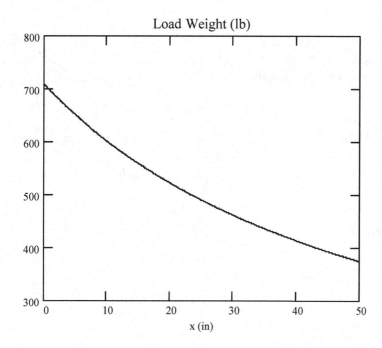

Load Weight (lb)

3.2 Problem 3/37 (2D Equilibrium)

The uniform 18-kg bar OA is held in the position shown by the smooth pin at O and the cable AB. Determine the tension T in the cable and the magnitude of the external pin reaction at O in terms of the angle θ. Plot the forces as a function of θ for $15^0 \leq \theta \leq 90^0$.

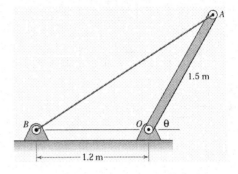

Problem Formulation

The free-body diagram for the bar is shown to the right. Before proceeding to the equilibrium equations we must first find the angle α in terms of θ. From the law of cosines and law of sines,

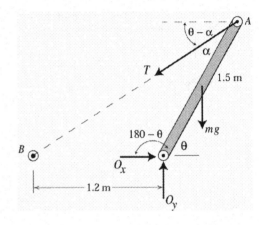

$$AB = \sqrt{1.2^2 + 1.5^2 - 2(1.2)(1.5)\cos(\pi - \theta)}$$

$$\frac{\sin\alpha}{1.2} = \frac{\sin(\pi - \theta)}{AB}$$

Observing that $\cos(\pi - \theta) = -\cos\theta$ and $\sin(\pi - \theta) = \sin\theta$,

$$AB = \sqrt{3.69 + 3.60\cos\theta}$$

$$\alpha = \sin^{-1}\left(\frac{1.2\sin\theta}{AB}\right)$$

The equilibrium equations can now be written,

$$\circlearrowleft \Sigma M_O = 0 = mg\left(\frac{1.5}{2}\cos\theta\right) - T\sin\alpha(1.5)$$

$$\Sigma F_x = 0 = O_x - T\cos(\theta - \alpha)$$

$$\Sigma F_y = 0 = O_y - mg - T\sin(\theta - \alpha)$$

The three equations above can be solved for the forces,

$$T = \frac{mg}{2}\frac{\cos\theta}{\sin\alpha}$$

$$O_x = T\cos(\theta - \alpha) \qquad\qquad O_y = mg + T\sin(\theta - \alpha)$$

$$O = \sqrt{O_x^2 + O_y^2}$$

MathCad Worksheet

m := 18   g := 9.81

$$AB(\theta) := \sqrt{3.69 + 3.6 \cdot \cos(\theta)}$$

$$\alpha(\theta) := \operatorname{asin}\left(\frac{1.2 \cdot \sin(\theta)}{AB(\theta)}\right)$$

$$T(\theta) := \frac{m \cdot g \cdot \cos(\theta)}{2 \cdot \sin(\alpha(\theta))}$$

$$O_X(\theta) := T(\theta) \cdot \cos(\theta - \alpha(\theta))$$

$$O_y(\theta) := m \cdot g + T(\theta) \cdot \sin(\theta - \alpha(\theta))$$

$$O(\theta) := \sqrt{O_X(\theta)^2 + O_y(\theta)^2}$$

x := 15, 15.1.. 90

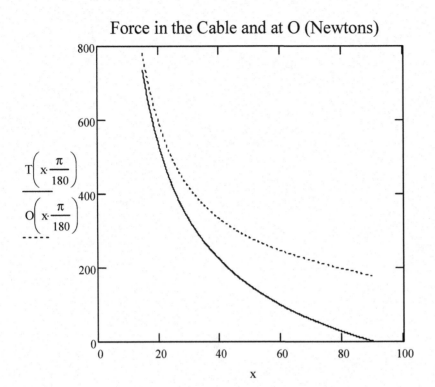

Force in the Cable and at O (Newtons)

Angle Theta (degrees)

3.3 Sample Problem 3/4 (2D Equilibrium)

Let the 10 kN load be located a distance c (meters) to the left of B. Find the magnitude T of the tension in the supporting cable and the magnitude of the force on the pin at A in terms of c. (a) Plot T and A as a function of c letting c range between 0 and 5 m. (b) Find the minimum and maximum values for T and A when c varies between 0 and 5 m. The beam AB is a standard 0.5-m I-beam with a mass of 95 kg per meter of length.

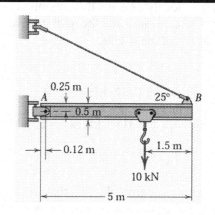

Problem Formulation

The free-body diagram is shown in the figure to the right. The weight of the beam is $(95)(5)(9.81) = 4660$ N or 4.66 kN. The weight acts at the center of the beam. The equilibrium equations are now written from the free-body diagram.

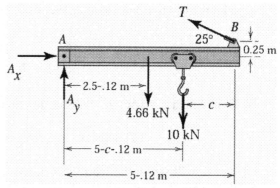

$$[\Sigma M_A = 0] \quad (T\cos 25°)0.25 + (T\sin 25°)(5 - 0.12)$$
$$- 10(5 - c - 0.12) - 4.66(2.5 - 0.12) = 0$$

from which $T = 26.15 - 4.367c$

$$[\Sigma F_x = 0] \quad A_x - T\cos 25° = 0$$
$$A_x = (26.15 - 4.367c)\cos 25° = 23.70 - 3.958c$$

$$[\Sigma F_y = 0] \quad A_y + T\sin 25° - 4.66 - 10 = 0$$
$$A_y = 5.66 - (26.15 - 4.367c)\sin 25° = 3.61 + 1.846c$$

$$A = \sqrt{A_x^2 + A_y^2} = \sqrt{(23.7 - 3.958c)^2 + (3.61 + 1.846c)^2}$$
$$A = \sqrt{574.7 - 174.3c + 19.07c^2}$$

(a) The plot of T and A as a function of c can be found in the worksheet below.

(b) The locations c for the maximum values of T and A and the minimum value for T are clear from the plot below. Substituting these values of c into the equations above yields,

$$T_{max} = 26.15 \text{ kN at } c = 0$$

$$T_{min} = 4.315 \quad \text{kN at } c = 5 \text{ m}$$

$$A_{max} = 23.97 \text{ kN at } c = 0$$

It is also clear from the plot that A goes through a minimum somewhere between $c = 4$ and 5 m. Exactly where this occurs can be determined by differentiating A with respect to c and equating the result to zero.

$$\frac{dA}{dc} = \frac{19.07c - 87.15}{\sqrt{574.7 - 174.3c + 19.07c^2}} = 0$$

from which $c = 4.569$ m. Substituting this value into A yields,

$$A_{min} = 13.286 \text{ kN at } c = 4.569 \text{ m}$$

As we will see in the worksheet below, Mathcad can be used to obtain many of the detailed aspects of the solution outlined above.

Mathcad Worksheet

$$\theta := 25 \cdot \frac{\pi}{180}$$

$$T(c) := 26.15 - 4.367 \cdot c$$

$$A_x(c) := T(c) \cdot \cos(\theta) \qquad A_y(c) := 14.66 - T(c) \cdot \sin(\theta)$$

$$A(c) := \sqrt{A_x(c)^2 + A_y(c)^2}$$

$c := 0, 0.05 .. 5$

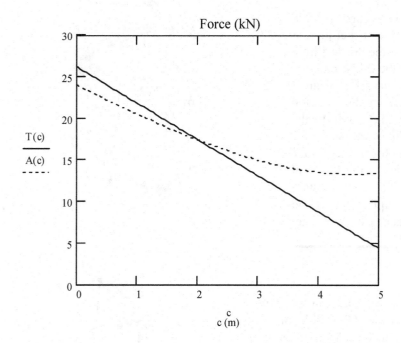

The locations for the maximum values of T and A and the minimum value for T are clear from the plot. We can obtain the associated maximum and minimum values by substitution.

$T_{max} := T(0)$ $T_{max} = 26.15$

$T_{min} := T(5)$ $T_{min} = 4.315$

$A_{max} := A(0)$ $A_{max} = 23.973$

To find the location of Amin we first differentiate A with respect to c and set the result equal to zero. The value of c determined by solving this equation is then substituted back into A to give Amin.

$dAdc(c) := \dfrac{d}{dc} A(c)$

$dAdc(x) = 0 \ solve, x \ \rightarrow 4.5693648460914463626$

$A_{min} := A(4.569364)$ $A_{min} = 13.286$

3.4 Sample Problem 3/7 (3D Equilibrium)

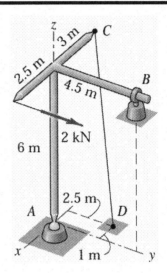

The welded tubular frame is secured to the horizontal x-y plane by a ball and socket joint at A and receives support from the loose-fitting ring at B. Under the action of the 2-kN load, the cable CD prevents rotation about a line from A to B, and the frame is stable in the position shown. Let the distance from the z-axis to the ring at B be c meters so that the coordinates of B are (0, c, 6). Find expressions for the tension T in the cable and the magnitudes of the forces at A and B in terms of c. Plot T, A, and B as a function of c for $0 \leq c \leq 6$ m. Explain why A and B go to infinity as c approaches 0. You may neglect the weight of the frame.

Problem Formulation

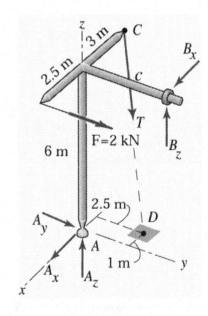

In the solution to the sample problem in your text, moments were summed about the AB axis in order to obtain one equation with one unknown, the tension T. When using computer software capable of symbolic algebra such considerations are far less important. In fact, it is often advisable to take the most straightforward approach in setting up the problem and then use the computer to work out the algebra.

Summing moments about A in the free-body diagram shown to the right,

$$[\Sigma M_A = 0] \qquad \mathbf{r}_{AD} \times \mathbf{T} + \mathbf{r}_{AE} \times \mathbf{F} + \mathbf{r}_{AB} \times \mathbf{B} = 0$$

where $\mathbf{T} = \dfrac{T(2\mathbf{i} + 2.5\mathbf{j} - 6\mathbf{k})}{\sqrt{(2)^2 + (2.5)^2 + (-6)^2}} = \dfrac{2T}{\sqrt{185}}(2\mathbf{i} + 2.5\mathbf{j} - 6\mathbf{k})$

$\mathbf{F} = 2\mathbf{j}$ (kN) $\mathbf{B} = B_x \mathbf{i} + B_Z \mathbf{k}$

$\mathbf{r}_{AD} = -\mathbf{i} + 2.5\mathbf{j}$ $\mathbf{r}_{AE} = 2.5\mathbf{i} + 6\mathbf{k}$ $\mathbf{r}_{AB} = c\mathbf{j} + 6\mathbf{k}$

Completion of the vector operations above yields a vector equation whose x, y, and z components give three scalar equations for the summation of moments about the x, y, and z-axes respectively,

$$\Sigma M_x = -\frac{6\sqrt{185}}{37}T - 12 + cB_z = 0 \qquad \Sigma M_y = -\frac{12\sqrt{185}}{185}T + 6B_x = 0$$

$$\Sigma M_z = -\frac{3\sqrt{185}}{37}T + 5 - cB_x = 0$$

These three equations can be solved simultaneously to give

$$T = \frac{5\sqrt{185}}{15 + 2c} \qquad B_x = \frac{10}{15 + 2c} \qquad B_z = \frac{6(55 + 4c)}{c(15 + 2c)}$$

The remaining unknowns can be found by summing forces,

$$\Sigma F_x = T_x + B_x + A_x = 0; \quad \Sigma F_y = T_y + 2 + A_y = 0;$$

$$\Sigma F_z = T_z + B_z + A_z = 0$$

Substituting the components of **T** (T_x, T_y, T_z) and **B** $(B_x \ B_z)$ into the above yields,

$$A_x = -\frac{30}{15 + 2c} \qquad A_y = -\frac{55 + 4c}{15 + 2c} \qquad A_z = \frac{6(6c - 55)}{c(15 + 2c)}$$

Finally, $\quad A = \sqrt{A_x^2 + A_y^2 + A_z^2} \quad$ and $B = \sqrt{B_x^2 + B_z^2}$

There are two ways to solve this problem with Mathcad. One is to go through the manipulations outlined above by hand and then use Mathcad to plot the results. Here we will use a second approach that makes better use of Mathcad's symbolic abilities. We will let Mathcad do the vector algebra (primarily cross products), which result in the equations summarized above. We will then use Mathcad to solve those equations. This results in a rather lengthy worksheet but avoids some tedious algebra.

Mathcad Worksheet

First we set up the three position vectors.

$$r_{ad} := \begin{pmatrix} -1 \\ 2.5 \\ 0 \end{pmatrix} \qquad r_{ae} := \begin{pmatrix} 2.5 \\ 0 \\ 6 \end{pmatrix} \qquad r_{ab} := \begin{pmatrix} 0 \\ c \\ 6 \end{pmatrix}$$

Note that c will appear in red on your worksheet since it has not been defined. Failure to define a variable will obviously affect numerical calculations but it will not affect the symbolic calculations below.

Now we want to express the tension as a Cartesian vector. We need to be careful to distinguish between the vector and its magnitude. Here we will define T_{cd} as the vector with magnitude T_m. We first define the position vector giving the direction of T and then define T as the magnitude times a unit vector found by dividing the position vector by its magnitude.

$$r_{cd} := \begin{pmatrix} 2 \\ 2.5 \\ -6 \end{pmatrix}$$

$$T_{cd} := T_m \cdot \frac{r_{cd}}{|r_{cd}|}$$

Now we define the force F and the reactions at A and B as vectors,

$$F := \begin{pmatrix} 0 \\ 2 \\ 0 \end{pmatrix} \qquad B := \begin{pmatrix} B_x \\ 0 \\ B_z \end{pmatrix} \qquad A := \begin{pmatrix} A_x \\ A_y \\ A_z \end{pmatrix}$$

Finally, we can find the expression for the summation of moments about *A* by carrying out the cross-products (see the problem formulation section).

$$r_{ad} \times T_{cd} + r_{ae} \times F + r_{ab} \times B \rightarrow \begin{pmatrix} -2.2056438662814232452 \cdot T_m - 12 + c \cdot B_z \\ -.88225754651256929808 \cdot T_m + 6 \cdot B_x \\ -1.1028219331407116226 \cdot T_m + 5.0 - c \cdot B_x \end{pmatrix}$$

The expression above evaluates the cross products for the summation of moments about A. This is a vector equation where the first, second and third terms in the bracketed expression correspond to our x, y and z-axes. From this we can get three scalar equations for the summation of moments about the x, y and z-axes. First we copy and paste the three equations, assigning each a name. We then solve the three expressions simultaneously for the magnitude of $\mathbf{T}$ and the x and z components of $\mathbf{B}$.

$$\text{sumMx} := -2.2056438662814232452 \cdot T_m - 12 + c \cdot B_z$$

$$\text{sumMy} := -.88225754651256929808 \cdot T_m + 6 \cdot B_x$$

$$\text{sumMz} := -1.1028219331407116226 \cdot T_m + 5.0 - c \cdot B_x$$

Given

$$\text{sumMx} = 0 \qquad \text{sumMy} = 0 \qquad \text{sumMz} = 0$$

$$\text{Find}(T_m, B_x, B_z) \rightarrow \begin{bmatrix} \dfrac{68.007352543677216723}{(15. + 2. \cdot c)} \\ \dfrac{10.}{(15. + 2. \cdot c)} \\ 6. \cdot \dfrac{(55. + 4. \cdot c)}{[c \cdot (15. + 2. \cdot c)]} \end{bmatrix}$$

From the above results we can now define the three forces as functions of c. This is necessary for the plot generated below.

$$T_m(c) := \frac{68.007352543677216723}{(15. + 2. \cdot c)}$$

$$B_x(c) := \frac{10.}{(15. + 2. \cdot c)}$$

$$B_z(c) := 6. \cdot \frac{(55. + 4. \cdot c)}{[c \cdot (15. + 2. \cdot c)]}$$

At this point we already have all forces expressed as vectors. We can thus write three scalar equations for the summation of forces in the x, y, and z directions by taking the first, second, and third terms in each vector. This is accomplished in Mathcad by using indices. Remember, though, that indices start at 0 in Mathcad. Thus, the x, y, and z components correspond to an index of 0, 1, and 2 respectively. To obtain an index you type "[". For example, the y component of B is evaluated by typing "B[1".

$$\text{sumFx} := T_{cd_0} + F_0 + B_0 + A_0$$

$$\text{sumFy} := T_{cd_1} + F_1 + B_1 + A_1$$

$$\text{sumFz} := T_{cd_2} + F_2 + B_2 + A_2$$

Given

$$\text{sumFx} \equiv 0 \qquad \text{sumFy} \equiv 0 \qquad \text{sumFz} \equiv 0$$

$$\text{Find}(A_x, A_y, A_z) \rightarrow \begin{pmatrix} -.29408584883752309936 \cdot T_m - 1. \cdot B_x \\ -.36760731104690387420 \cdot T_m - 2. \\ .88225754651256929808 \cdot T_m - 1. \cdot B_z \end{pmatrix}$$

From the above we can now express the three components of the force at *A* as functions of *c*.

$$A_x(c) := -.29408584883752309936 \cdot T_m(c) - 1. \cdot B_x(c)$$

$$A_y(c) := -.36760731104690387420 \cdot T_m(c) - 2.$$

$$A_z(c) := .88225754651256929808 \cdot T_m(c) - 1. \cdot B_z(c)$$

$$A(c) := \sqrt{A_x(c)^2 + A_y(c)^2 + A_z(c)^2} \qquad B(c) := \sqrt{B_x(c)^2 + B_z(c)^2}$$

$c := 0, .05 .. 10$

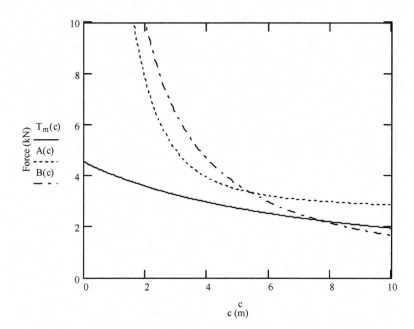

Note that the ring can no longer support the frame when it is located exactly on the *z*-axis. It is for this reason that the forces *A* and *B* go to infinity as *c* approaches 0.

STRUCTURES

4

This chapter concerns the determination of internal forces in a structure. A structure is an assembly of connected members designed to support or transfer forces. Thus, the internal forces that are generally of interest are the forces of action and reaction between the connected members of the structure. The types of structures considered here are generally classified as trusses, frames or machines.

Problems 4.1 and 4.2 use the methods of joints and sections respectively to analyze a two dimensional truss. In problem 4.2, the algebra is done "by hand", however, problem 4.1 utilizes the symbolic *Given...Find* to solve two equations for two unknown forces. Problem 4.3 uses the method of joints on a space truss. Once again, *Given...Find* is used to solve three equations symbolically. Problems 4.4 and 4.5 cover frames and machines.

4.1 Problem 4/14 (Trusses. Method of Joints)

The truss is composed of equilateral triangles of sides a and is loaded and supported as shown. Determine the reaction force on the roller at E and the forces in members FE, and DE as a function of θ. Plot the non-dimensional loads E/L, FE/L, and DE/L for θ between 0 and 45 degrees.

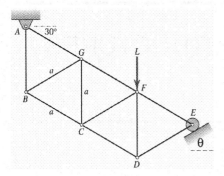

Problem Formulation

First we determine the reaction force at E from a free-body diagram for the entire truss (shown to the right).

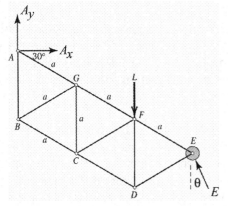

$$\Sigma M_A = -L(2a\cos(30)) + E\cos\theta(3a\cos(30))$$
$$- E\sin\theta(3a\sin(30)) = 0$$

$$E = \frac{2}{3}\frac{L\cos 30}{\cos\theta\cos 30 - \sin\theta\sin 30} = \frac{2}{3}\frac{L\sqrt{3}}{\sqrt{3}\cos\theta - \sin\theta}$$

Note that the non-dimensional force E/L can be easily found by dividing both sides of the equation by L.

$$\frac{E}{L} = \frac{2}{3}\frac{\sqrt{3}}{\sqrt{3}\cos\theta - \sin\theta}$$

Note also that this same equation could be obtained just as easily by setting $L = 1$. Thus one interpretation of a non-dimensional load such as E/L is the force per unit load L.

To obtain the required forces FE and DE we now consider a free-body diagram for joint E. Note that each member is assumed to be in tension. Thus, positive answers will imply tension and negative answers compression.

Joint E

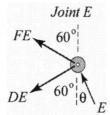

Joint E

$$\Sigma F_x = 0 = -FE\sin 60 - DE\sin 60 - E\sin\theta$$

$$\Sigma F_y = 0 = FE\cos 60 - DE\cos 60 + E\cos\theta$$

Note that after E is substituted that the above are two equations in two unknowns for FE and DE. We will let MathCad carry out the solution to these equations in the worksheet below. The result is

$$\frac{FE}{L} = -\frac{2}{3}\frac{\sqrt{3}\cos\theta + \sin\theta}{\sqrt{3}\cos\theta - \sin\theta} \qquad\qquad \frac{DE}{L} = \frac{2}{3}$$

Of course, the fact that the force in member DE is 2L/3 independent of θ is probably surprising.

MathCad Worksheet

First, solve the equilibrium equations from joint E. Note how the result for the reaction at E is included with the **Given** so that it does not have to be explicitly substituted into each equation.

Given

$$E = \frac{2}{3}\cdot\frac{\sqrt{3}}{\cos(\theta)\cdot\sqrt{3} - \sin(\theta)}$$

$$FE\cdot\sin\left(60\cdot\frac{\pi}{180}\right) + DE\cdot\sin\left(60\cdot\frac{\pi}{180}\right) + E\cdot\sin(\theta) = 0$$

$$FE\cdot\cos\left(60\cdot\frac{\pi}{180}\right) - DE\cdot\cos\left(60\cdot\frac{\pi}{180}\right) + E\cdot\cos(\theta) = 0$$

$$\text{Find}(FE, DE, E) \rightarrow \begin{pmatrix} \dfrac{-2}{3} \cdot \dfrac{\cos(\theta) \cdot 3^{\frac{1}{2}} + \sin(\theta)}{\cos(\theta) \cdot 3^{\frac{1}{2}} - \sin(\theta)} \\[2em] \dfrac{2}{3} \\[2em] \dfrac{2}{3} \cdot \dfrac{3^{\frac{1}{2}}}{\cos(\theta) \cdot 3^{\frac{1}{2}} - \sin(\theta)} \end{pmatrix}$$

Copying and pasting from above:

$$DE := \frac{2}{3}$$

$$E(\theta) := \frac{2}{3} \cdot \frac{3^{\frac{1}{2}}}{3^{\frac{1}{2}} \cdot \cos(\theta) - \sin(\theta)}$$

$$FE(\theta) := \frac{-2}{3} \cdot \frac{3^{\frac{1}{2}} \cdot \cos(\theta) + \sin(\theta)}{3^{\frac{1}{2}} \cdot \cos(\theta) - \sin(\theta)}$$

$$\theta := 0, 0.01 .. \frac{\pi}{4}$$

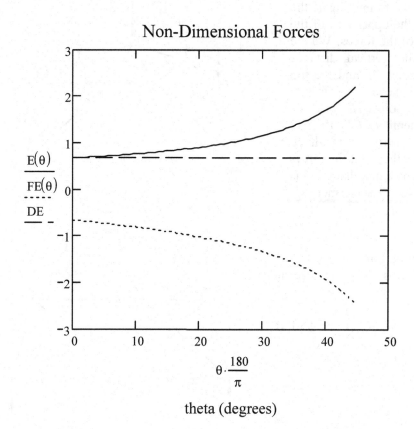

Non-Dimensional Forces

$$\theta \cdot \frac{180}{\pi}$$

theta (degrees)

4.2 Problem 4/48 (Trusses. Method of Sections)

In this problem we would like to investigate the effects of the geometry of the upper part of the arched roof truss on some of the forces. We are given a constraint that the vertical distance between the horizontal section BF and the top most point must be 7 m, however we can vary the vertical location of points E and C. Determine the forces in members DE, EI, EF, FI, and HI of the truss. First, write the results in terms of the lengths a and b, then plot the forces as a function of a with the constraint that $a + b$ is always 7 m. Let a vary between 1 and 6 meters.

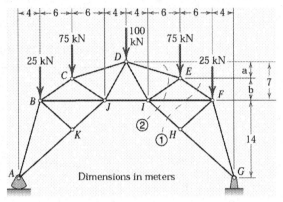

Problem Formulation

First we determine the reaction force at G from a free-body diagram for the entire truss (shown to the right). Due to symmetry,

$$A_y = G_y = (25 + 75 + 100 + 75 + 25)/2 = 150 \text{ kN}$$

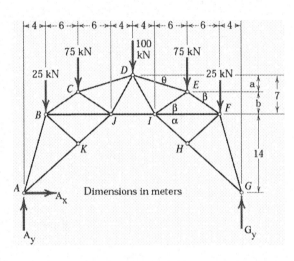

Dimensions in meters

From the free-body diagram we also find the angles,

$$\theta = \tan^{-1}\left(\frac{a}{6}\right) \quad \beta = \tan^{-1}\left(\frac{b}{6}\right) \quad \alpha = \tan^{-1}\left(\frac{14}{16}\right)$$

From the free-body diagram for section 1

$$\circlearrowleft \Sigma M_I = 0 = 150(16) - 25(12) + EF \sin\beta(12)$$

$$EF = -\frac{175}{\sin\beta}$$

$$\circlearrowleft \Sigma M_F = 0 = 150(4) - IH \sin\alpha(12)$$

$$IH = \frac{50}{\sin\alpha}$$

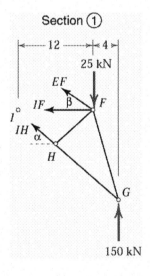

Section ①

$$\Sigma F_x = 0 = -IF - EF \cos \beta - IH \cos \alpha$$

$$IF = -EF \cos \beta - IH \cos \alpha = \frac{175}{\tan \beta} - \frac{50}{\tan \alpha}$$

Note that the summation of moments about F was simplified by sliding IH along its line of action to I before resolving it into its horizontal and vertical components.

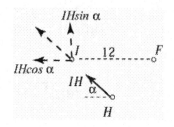

Now look at the free-body diagram for section 2. Forces IF and IH are already known, so we need only two equilibrium equations.

$$\Sigma M_I = 0 = DE \cos \theta (b) + DE \sin \theta (6)$$
$$- 75(6) - 25(12) + 150(16)$$

$$DE = \frac{-1650}{6 \sin \theta + b \cos \theta}$$

$$\Sigma F_x = 0 = IH \cos \alpha + IF + EI \cos \beta + DE \cos \theta$$

$$EI = -\frac{IH \cos \alpha + IF + DE \cos \theta}{\cos \beta}$$

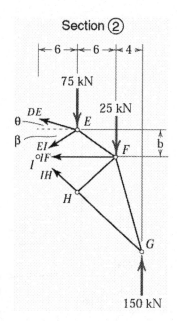

We will allow the computer to make the substitutions of IH, IF, and DE in the last equation. Note that all the members were assumed to be in tension. Thus positive results will imply tension and negative results compression.

MathCad Worksheet

$b(a) := 7 - a$

$$\theta(a) := \text{atan}\left(\frac{a}{6}\right) \qquad \beta(a) := \text{atan}\left(\frac{b(a)}{6}\right) \qquad \alpha := \text{atan}\left(\frac{14}{16}\right)$$

$$EF(a) := \frac{-175}{\sin(\beta(a))} \qquad\qquad IH(a) := \frac{50}{\sin(\alpha)}$$

$$IF(a) := \frac{175}{\tan(\beta(a))} - \frac{50}{\tan(\alpha)}$$

$$DE(a) := \frac{-1650}{6 \cdot \sin(\theta(a)) + b(a) \cdot \cos(\theta(a))}$$

$$EI(a) := \frac{-(IH(a) \cdot \cos(\alpha) + IF(a) + DE(a) \cdot \cos(\theta(a)))}{\cos(\beta(a))}$$

$$a := 1, 1.01 .. 6$$

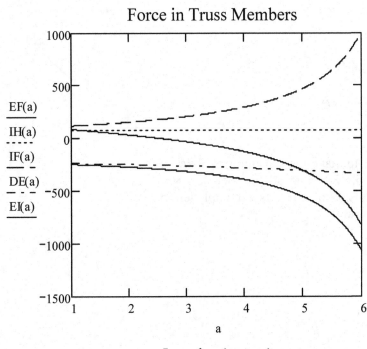

Length a (meters)

4.3 Sample Problem 4/5 (Space Trusses)

The space truss consists of the rigid tetrahedron *ABCD* anchored by a ball-and-socket connection at *A* and prevented from any rotation about the *x*-, *y*-, or *z*-axes by the respective links 1, 2, and 3. The load *L* is applied at joint *E*, which is rigidly fixed to the tetrahedron by the three additional links. Here we would like to investigate how the forces in a few of the members depend upon the height of the structure so let the height be *d* instead of 4 meters. Solve for the forces in the members at joint *E* and plot these as a function of *d* if *L* = 10 kN. Let *d* range between 2 and 10 meters.

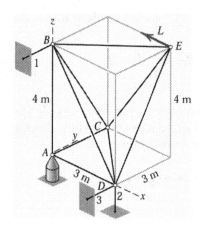

Problem Formulation

We could begin by analyzing a free-body diagram for the entire structure; however, this is not necessary in the present case as joint *E* contains only three unknown forces that happen to be the ones we are interested in. Joint *E* also has the external force *L* so that we can solve this problem from a single free-body diagram for joint *E*, shown to the right. Refer to the solution to this problem in your text for an explanation of the general procedure for determining the forces in all of the members. Our first step is to express the forces at *E* as Cartesian vectors.

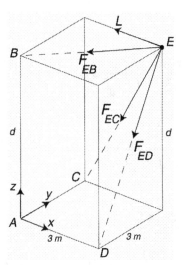

$$\mathbf{F_{EC}} = F_{EC}\frac{-3\mathbf{i}-d\mathbf{k}}{\sqrt{9+d^2}} \qquad \mathbf{F_{ED}} = F_{ED}\frac{-3\mathbf{j}-d\mathbf{k}}{\sqrt{9+d^2}}$$

$$\mathbf{F_{EB}} = \frac{F_{EB}}{\sqrt{2}}(-\mathbf{i}-\mathbf{j}) \qquad \mathbf{L} = -L\mathbf{i}$$

The scalar equilibrium equations can now be written from the *x*-, *y*-, and *z*-components of the vector equations above.

$$\Sigma F_x = 0 = -L - \frac{F_{EB}}{\sqrt{2}} - \frac{3F_{EC}}{\sqrt{9+d^2}} \qquad \Sigma F_y = 0 = -\frac{F_{EB}}{\sqrt{2}} - \frac{3F_{ED}}{\sqrt{9+d^2}}$$

$$\Sigma F_z = 0 = -\frac{d}{\sqrt{9+d^2}}\left(F_{EC} + F_{ED}\right)$$

These equations can be solved for the three forces,

$$F_{ED} = -F_{EC} = \frac{L}{6}\sqrt{9+d^2} \qquad F_{EB} - \frac{L}{\sqrt{2}}$$

Referring back to the free-body diagram we see that all three members were assumed to be in tension. Thus, the results indicate that *ED* is in tension while *EC* and *EB* are in compression.

Mathcad Worksheet

First we use symbolic algebra to solve the equilibrium equations for the unknown forces.

Given

.

$$-L - \frac{F_{EB}}{\sqrt{2}} - \frac{3 \cdot F_{EC}}{\sqrt{9+d^2}} \equiv 0$$

$$\frac{-F_{EB}}{\sqrt{2}} - \frac{3 \cdot F_{ED}}{\sqrt{9+d^2}} \equiv 0$$

$$F_{EC} + F_{ED} \equiv 0$$

$$\text{Find}\left(F_{EC}, F_{ED}, F_{EB}\right) \rightarrow \begin{bmatrix} \frac{-1}{6} \cdot L \cdot \left(9+d^2\right)^{\left(\frac{1}{2}\right)} \\ \frac{1}{6} \cdot L \cdot \left(9+d^2\right)^{\left(\frac{1}{2}\right)} \\ \frac{-1}{2} \cdot L \cdot \sqrt{2} \end{bmatrix}$$

$L := 10$

$$F_{EB} := \frac{-1}{2} \cdot L \cdot \sqrt{2} \qquad\qquad F_{ED}(d) := \frac{1}{6} \cdot L \cdot \left(9 + d^2\right)^{\left(\frac{1}{2}\right)}$$

$$F_{EC}(d) := -F_{ED}(d)$$

$d := 2, 2.05 .. 10$

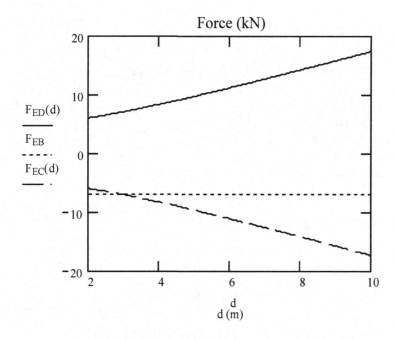

Force (kN)

$F_{ED}(d)$
——
F_{EB}
- - - -
$F_{EC}(d)$
— .

d
d (m)

4.4 Problem 4/83 (Frames and Machines)

Determine the magnitude of the pin reaction at A and the magnitude of the force reaction at the rollers as a function of x. The pulleys at C and D are small. Plot the two forces for $0 \leq x \leq 0.8$ m. What is the direction of the force at the rollers? Does the direction change for the range of x considered?

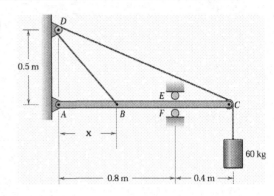

Problem Formulation

$$\alpha = \tan^{-1}(0.5/x) \qquad \theta = \tan^{-1}(0.5/1.2) = 22.62°$$

$$\circlearrowleft M_A = 0 = Tx\sin\alpha + T(1.2)\sin\theta - T(1.2) + EF(0.8)$$

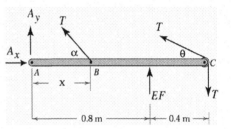

Note in the Free Body Diagram that we have assumed that force EF acts up. This is equivalent to assuming that the boom is in contact with and pushing down on roller F. Roller F in turn pushes back up on the boom. This assumption is verified by calculating only positive values for the force.

Substituting $\theta = 22.62°$, T = 60(9.81), and solving gives,

$$EF = 543.32 - 735.75x\sin\alpha$$

$$\Sigma F_x = 0 = A_x - T\cos\theta - T\cos\alpha \qquad A_x = T(\cos\theta + \cos\alpha)$$

$$\Sigma F_y = 0 = A_y + T\sin\alpha + T\sin\theta + EF - T \qquad A_y = T(1 - \sin\alpha - \sin\theta) - EF$$

$$A = \sqrt{A_x^2 + A_y^2}$$

The formulation is now complete since the tow forces EF and A are known as a function of x. Note once again that we do not have to make explicit substitutions. For example, EF is written in terms of x and α, but α is already known as a function of x.

Mathcad Worksheet

$$\alpha(x) := \text{atan}\left(\frac{.5}{x}\right) \quad \theta := \text{atan}\left(\frac{.5}{1.2}\right)$$

Tension := 60·9.81

$$EF(x) := 543.32 - 735.75 \cdot x \cdot \sin(\alpha(x))$$

$$Ax(x) := \text{Tension} \cdot (\cos(\theta) + \cos(\alpha(x)))$$

$$Ay(x) := \text{Tension} \cdot (1 - \sin(\alpha(x)) - \sin(\theta)) - EF(x)$$

$$A(x) := \sqrt{Ax(x)^2 + Ay(x)^2}$$

$$x := 0, 0.01 .. 0.8$$

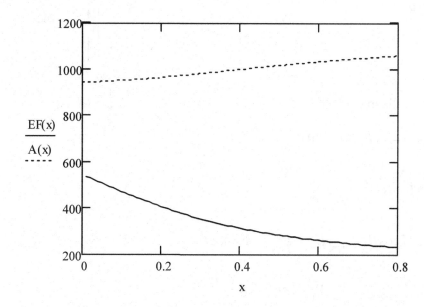

4.5 Problem 4/108 (Frames and Machines)

A lifting device for transporting 135-kg steel drums is shown. Develop expressions for the forces at A, C, and E in terms of L, the length of member AG. Plot these forces as a function L between 345 and 380 mm.

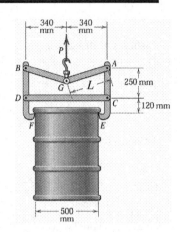

Problem Formulation

$$\theta = \cos^{-1}(340/L) \qquad P = W = 135(9.81)\ \text{N}$$

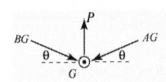

Since BG and AG are two-force members we can draw a free-body of pin G as shown to the right. By symmetry, $AG = BG$, so

$$\Sigma F_y = 0 = P - 2AG\sin\theta$$

$$AG = \frac{P}{2\sin\theta}$$

Now consider the free-body diagram of ACE.

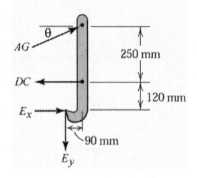

$$\circlearrowleft M_E = 0 = AG\sin\theta(90) - AG\cos\theta(370) + DC(120)$$

$$DC = \left(\frac{37}{12}\cos\theta - \frac{3}{4}\sin\theta\right)AG$$

$$\Sigma F_x = 0 = E_x - DC + AG\cos\theta \qquad E_x = DC - AG\cos\theta$$

$$\Sigma F_y = 0 = AG\sin\theta - E_y \qquad E_y = AG\sin\theta = \frac{P}{2}$$

$$E = \sqrt{E_x^2 + E_y^2}$$

The formulation is now complete since all forces are known in terms of θ which is in turn known as a function of L.

MathCad Worksheet

$P := 135 \cdot 9.81$

$$\theta(L) := \text{acos}\left(\frac{340}{L}\right)$$

$$AG(L) := \frac{P}{2 \cdot \sin(\theta(L))}$$

$$DC(L) := AG(L) \cdot \left(\frac{37 \cdot \cos(\theta(L))}{12} - \frac{3}{4} \cdot \sin(\theta(L))\right)$$

$$E_x(L) := DC(L) - AG(L) \cdot \cos(\theta(L))$$

$$E_y(L) := AG(L) \cdot \sin(\theta(L))$$

$$E(L) := \sqrt{E_x(L)^2 + E_y(L)^2}$$

L := 345, 345.1 .. 380

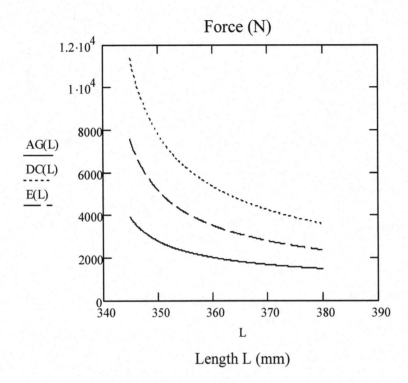

Force (N)

AG(L)
——
DC(L)

E(L)
— —

L

Length L (mm)

DISTRIBUTED FORCES

5

This chapter considers forces that are distributed over lines, areas, or volumes. Also considered are a few preliminary topics directly related to distributed forces such as center of mass and centroids. The major applications are beams, cables and fluid statics.

Problem 5.1 contains the first example in this booklet of a parametric plot. Problem 5.2 formulates the equilibrium equations for a beam in terms of integrals of the distributed load on the beam. Problem 5.3 expresses the internal shear force and bending moment in terms of the external loading function $w(x)$. The integrals are evaluated symbolically in MathCad. *root* is used to obtain a numerical solution for the minimum tension and total length of a flexible cable in problem 5.4. Problem 5.5 is a fluid statics problem.

5.1 Sample Problem 5/8 (Composite Bodies)

Let the length of the 40-mm diameter shaft in the bracket-and-shaft assembly be L (mm). The vertical face of the assembly is made from sheet metal that has a mass of 25 kg/m^2. The material of the horizontal base has a mass of 40 kg/m^2, and the steel shaft has a density of 7.83 Mg/m^3. (a) Find expressions for the location of the center of mass ($\bar{Y}$, $\bar{Z}$) in terms of L. (b) Plot $\bar{Y}$ and $\bar{Z}$ as a function of L for $0 \leq L \leq 200$ mm. (c) Plot $\bar{Z}$ versus $\bar{Y}$ for $0 \leq L \leq 200$ mm.

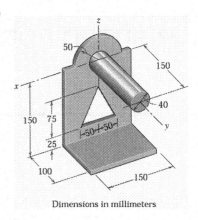

Dimensions in millimeters

Problem Formulation

(a) Refer to the solution to this sample problem in your text and be sure that you understand how the results in the table were obtained. For the problem considered here, the table will remain unchanged except for the last row. For the steel shaft (part 5) we have,

$$m = 0.00984L; \quad \bar{y} = L/2; \quad \bar{z} = 0; \quad m\bar{y} = 0.00492L^2; \quad m\bar{z} = 0.$$

The totals become

$$\Sigma m = 1.166 + 0.00984L; \quad \Sigma m\bar{y} = 30 + 0.00492L^2; \quad \Sigma m\bar{z} = -120.73$$

Thus,

$$\bar{Y} = \frac{\Sigma m\bar{y}}{\Sigma m} = \frac{30 + 0.00492L^2}{1.166 + 0.00984L}; \quad \bar{Z} = \frac{\Sigma m\bar{z}}{\Sigma m} = \frac{-120.73}{1.166 + 0.00984L}$$

(b) The plot of $\bar{Y}$ and $\bar{Z}$ versus L can be found in the Mathcad worksheet below.

(c) At first sight, plotting $\bar{Z}$ versus $\bar{Y}$ may seem problematic since $\bar{Z}$ is not known explicitly as a function of $\bar{Y}$. Perhaps the most straightforward way of dealing with this apparent difficulty is to solve the first equation above for L in terms of $\bar{Y}$ and then substitute this result into the second equation. Though this approach might seem simple at first, it turns out that there are two solutions, both of which are very messy. There may also be situations where this substitution approach may be impossible, even with symbolic algebra.

Fortunately, most software packages allow plotting two variables without performing a substitution, provided the two variables are expressed in terms of a common parameter (in our case L). Plots of this type are usually called parametric plots. The parametric plot of $\bar{Z}$ versus $\bar{Y}$ can be found in the Mathcad worksheet below.

Mathcad Worksheet

$\text{sum\_mass}(L) := 1.166 + 0.00984L$

$$yb(L) := \frac{30 + 0.00492L^2}{\text{sum\_mass}(L)}$$

$$zb(L) := \frac{-120.73}{\text{sum\_mass}(L)}$$

$L := 0, 0.5 .. 200$

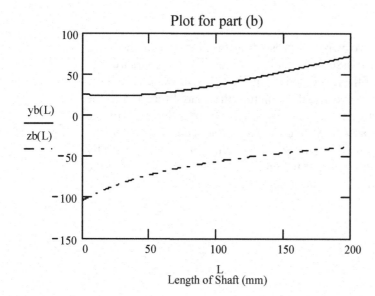

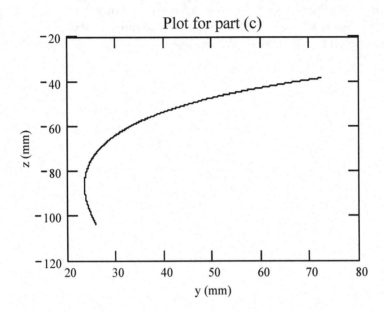

In this problem we have seen two ways of presenting the location of the mass center of the bracket-and-shaft assembly and it is useful to compare the advantages and disadvantages of these methods. The first plot shows clearly how $\overline{Y}$ and $\overline{Z}$ depend on L, however, it is difficult to picture the spatial location of the mass center. In the second plot one can actually see the spatial location of the mass center and how it moves as L is varied. What is not seen are the precise values of L at each combination of $\overline{Y}$ and $\overline{Z}$. This situation can be partially remedied by determining the values of L at the end-points of the curve. From the first plot you can easily verify that the curve begins (lower left corner) at $L = 0$ and ends at $L = 200$ mm.

5.2 Problem 5/113 (Beams-External Effects)

The load per foot of beam length varies as shown. Here we would like to look at the effects of the initial load w_0 on the reactions given the constraint that the total resultant R of the distributed load w remains constant. This means that we are, in effect, studying the effects of the shape of the load. First find the required value of k in terms of w_0 and L and then (a) plot the distributed load ($w(x)$) for w_0 = 100, 200, and 300 lb/ft and (b) plot the reactions at the two supports as a function of w_0 for $0 \le w_0 \le 300$ lb/ft. For (a) and (b) let R = 3500 lb and L = 20 ft.

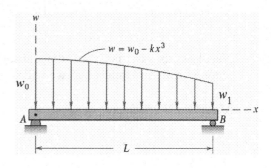

Problem Formulation

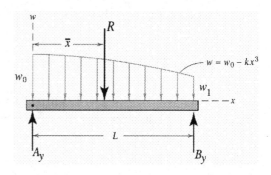

$$R = \int_0^L w\,dx = \int_0^L \left(w_0 - kx^3\right)dx = w_0 L - \frac{1}{4}kL^4$$

Solving,

$$k = \frac{4\left(w_0 L - R\right)}{L^4}$$

$$\bar{x} = \frac{1}{R}\int_0^L xw\,dx = \frac{1}{R}\int_0^L x\left(w_0 - kx^3\right)dx = \frac{1}{R}\left(\frac{1}{2}w_0 L^2 - \frac{1}{5}kL^5\right)$$

Substituting for k,

$$\bar{x} = \frac{L\left(8R - 3w_0 L\right)}{10R}$$

Now, from the free-body diagram,

$$\circlearrowleft \Sigma M_A = 0 = B_y L - R\bar{x}$$

$$\Sigma F_y = 0 = A_y + B_y - R$$

From which we find,

$$B_y = \frac{\bar{x}}{L} R \qquad A_y = R - B_y = \left(1 - \frac{\bar{x}}{L}\right) R$$

Needless to say, most of the substitutions indicated above are performed automatically in Mathcad, as we will see below.

MathCad Worksheet

$L := 20 \qquad R := 3500$

$$k\left(w_0\right) := \frac{4 \cdot \left(w_0 \cdot L - R\right)}{L^4}$$

$$w\left(x, w_0\right) := w_0 - k\left(w_0\right) \cdot x^3$$

$x := 0, .05 .. L$

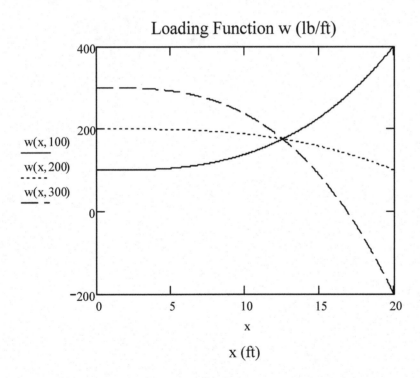

x (ft)

$$xbar(w_0) := \frac{L \cdot (8 \cdot R - 3 \cdot w_0 \cdot L)}{10 \cdot R}$$

$$B(w_0) := R \cdot \frac{xbar(w_0)}{L} \qquad A(w_0) := R - B(w_0)$$

$$w_0 := 0, .1 .. 300$$

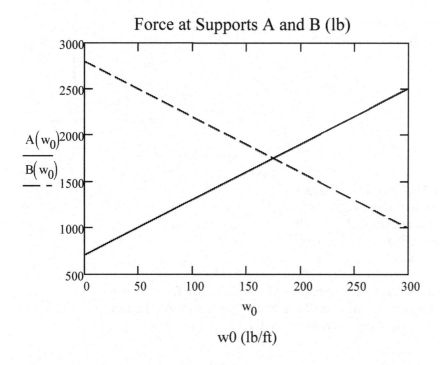

Force at Supports A and B (lb)

w0 (lb/ft)

5.3 Problem 5/140 (Beams-Internal Effects)

Derive expressions for the shear force V and bending moment M as functions of x for the cantilever beam loaded as shown. For the case where $L = 10$ ft and $w_L = 400$ lb/ft, (a) plot the distributed load $w(x)$, (b) the shear force $V(x)$ and (c) the bending moment $M(x)$. In each case plot three curves for $w_0 = -400$, 0, and 400 lb/ft.

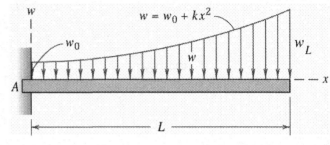

Problem Formulation

First, we need to express k in terms of w_0 and w_L. This is accomplished by substituting $x = L$ into the loading function,

$$w(x = L) = w_0 + kL^2 = w_L$$

$$k = \frac{w_L - w_0}{L^2}$$

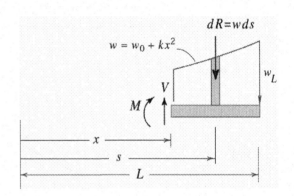

The easiest way to find $V(x)$ and $M(x)$ is from the free-body diagram shown to the right since we avoid having to find the reactions at the wall support. We do, however, have to be very careful setting up our integrals.

When we say $V(x)$, x is a coordinate measured positive to the right from the left end of the beam. It is essential to distinguish between that x and the dummy integration variable which will vary between x and L. Here, we denote the integration variable by s.

$$\Sigma F_y = V - \int dR \qquad V = \int_x^L w(s)\,ds = \int_x^L \left(w_0 + ks^2\right)ds$$

$$V(x) = w_0(L - x) + \frac{k}{3}\left(L^3 - x^3\right)$$

Before we sum moments first note that the moment arm of dR about a point on the cross-section at x is $s - x$,

$$\circlearrowleft \Sigma M = 0 = -M - \int (s-x)dR$$

$$M(x) = -\int_x^L (s-x)w(s)ds = -\int_x^L (s-x)(w_0 + ks^2)ds$$

$$M(x) = xw_0(L-x) - \frac{w_0}{2}(L^2 - x^2) + \frac{xk}{3}(L^3 - x^3) - \frac{k}{4}(L^4 - x^4)$$

MathCad Worksheet

$L := 10 \qquad w_L := 400$

$$k(w_0) := \frac{w_L - w_0}{L^2}$$

$$w(x, w_0) := w_0 + k(w_0) \cdot x^2$$

$x := 0, .05 .. L$

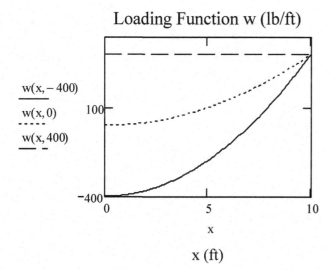

Loading Function w (lb/ft)

$\frac{w(x, -400)}{w(x, 0)}$

$w(x, 400)$

x (ft)

$$\underset{\sim}{V}(x, w_0) := \int_x^L w(s, w_0)\, ds$$

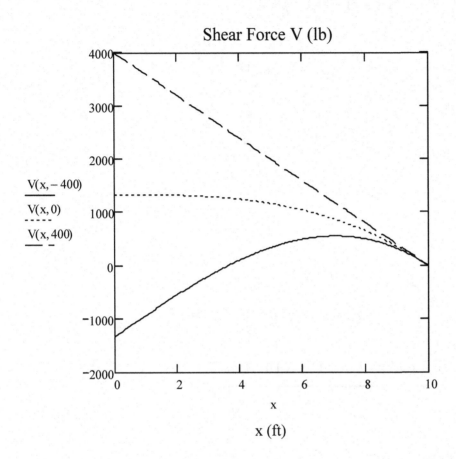

x (ft)

$$M(x, w_0) := -\int_x^L (s-x) \cdot (w(s, w_0)) \, ds$$

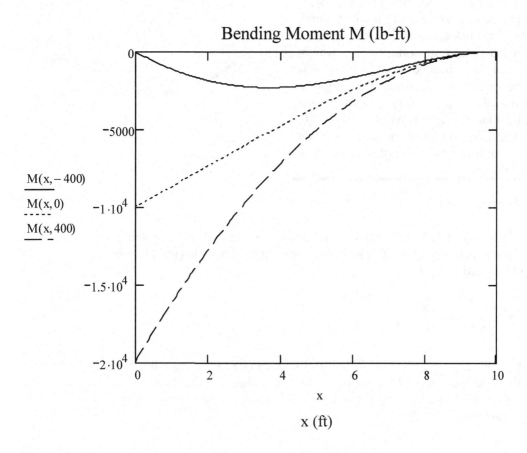

Bending Moment M (lb-ft)

$M(x, -400)$

$M(x, 0)$

$M(x, 400)$

x

x (ft)

To obtain the equations for the shear force or moment, use the symbolic ®

$V(s, 400) \rightarrow 4000 - 400 \cdot s$

$M(s, 400) \rightarrow (-20000) + 4000 \cdot s - 200 \cdot s^2$

5.4 Sample Problem 5/17 (Flexible Cables)

Replace the cable of Sample Problem 5/16, which is loaded uniformly along the horizontal, by a cable which has a mass of 12 kg per meter of its own length and supports its own weight only. The cable is suspended between two points on the same level 300 m apart and has a sag of h meters (instead of the 60 m shown in the diagram. (a) Plot y as a function of x for $h = 10$, 30, and 60 meters. (b) Plot the minimum tension T_0 and the total length ($L_c = 2s$) of the cable as a function of the sag h.

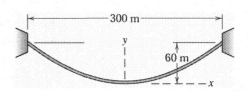

Problem Formulation

(a) For a uniformly distributed load we have a catenary shape for the cable as described in part (c) of Article 5/8. The curve assumed by the cable ($y(x)$) is thus described by Equation 5/19

$$y = \frac{T_0}{\mu}\left(\cosh \frac{\mu x}{T_0} - 1\right)$$

where $\mu = (12)(9.81)(10^{-3}) = 0.1177$ kN/m. Before y can be plotted as a function of x it is first necessary to find the value of T_0 corresponding to the three values of h that are given. As in the sample problem, this is accomplished by evaluating the above equation at $x = 150$ m,

$$h = \frac{T_0}{\mu}\left(\cosh \frac{150\mu}{T_0} - 1\right)$$

Since there are only three cases to consider it might seem reasonable to try the graphical method used in the sample problem, however, this approach is obviously too cumbersome for part (b). Therefore it is best to use a numerical approach. This is easily accomplished using Mathcad's *root* function. To use the *root* function we first need to re-write the above in terms of a function whose root (zero) provides the solution to the original equation.

$$f = h - \frac{T_0}{\mu}\left(\cosh \frac{150\mu}{T_0} - 1\right)$$

The roots ($f = 0$) of this equation are found for $h = 10$, 30, and 60 meters yielding $T_0 = 132.6$, 44.7, and 23.2 kN respectively. With T_0 known y can easily be plotted as a function of x by substituting into the equation above. The result is given in the worksheet below.

(b) It is possible to use the general approach for part (a) here as well. What you would need to do is set up a table of values for h and T_0. You would then use *root* to find T_0 for each value of h in the table. Finally, the tabular results would be entered into a graphics program (or a spreadsheet such as Excel) and the results plotted.

It turns out that you can automate the process with only a few statements in Mathcad. The general procedure is as follows (see the worksheet for details). First, a loop structure is set up which iterates on h. The Mathcad command *root* is placed within the loop to evaluate T_0 for each h. After T_0 has been determined, the total length of the cable ($L_c = 2s$) can be easily found from Equation 5/20 (with $x = 150$ m). The results are then plotted after the loop has completed.

Mathcad Worksheet

$L := 300 \quad \mu := 0.1177$

First we define the function f whose roots provide the solution to our equation.

$$f(T_0, h) := h - \frac{T_0}{\mu} \cdot \left(\cosh\left(\frac{\mu \cdot L}{2 \cdot T_0} \right) - 1 \right)$$

$T0 := 1$

This is the initial guess required when using the root function. It turns out that the same initial guess can be used for all values of h considered here. This is very convenient. In the following calculations, T0_10, T0_30, and T0_60 are the minimum cable tensions for h = 10, 30, and 60 meters.

$T0\_10 := \text{root}(f(T0, 10), T0) \qquad T0\_10 = 132.607$

$T0\_30 := \text{root}(f(T0, 30), T0) \qquad T0\_30 = 44.714$

$T0\_60 := \text{root}(f(T0, 60), T0) \qquad T0\_60 = 23.159$

$$y(x, T_0) := \frac{T_0}{\mu} \cdot \left(\cosh\left(\frac{\mu \cdot x}{T_0} \right) - 1 \right)$$

$x := -150, -149 .. 150$

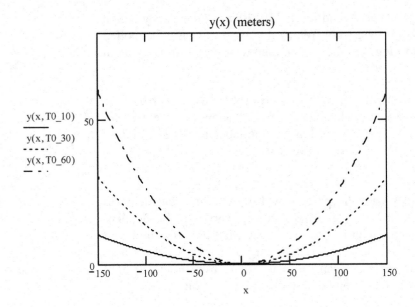

Plotting T0 versus h requires us to find T0 for a range of values for h. This is accomplished by essentially the same procedure as that used above except that we use a range variable to iterate on h.

$$i := 3, 4 .. 60 \qquad h_i := i$$

$$T0_i := \text{root}\big(f(T0, h_i), T0\big)$$

$$Lc_i := \frac{2 \cdot T0_i}{\mu} \cdot \sinh\left(\frac{\mu \cdot 150}{T0_i}\right)$$ Total length of the cable (from Equation 5/20)

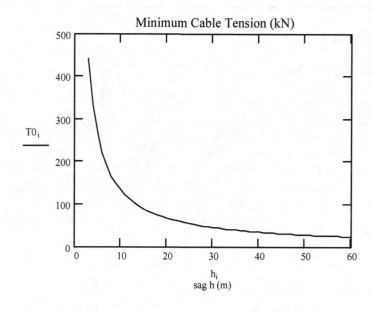

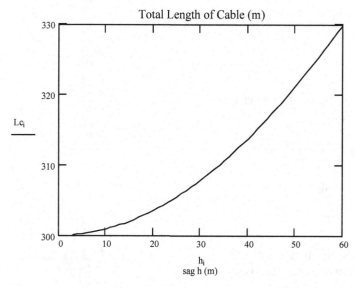

5.5 Problem 5/189 (Fluid Statics)

The rectangular gate shown in section is 10 ft long (perpendicular to the paper) and is hinged about its upper edge *B*. The gate divides a channel leading to a fresh-water lake on the left and a salt-water tidal basin on the right. Calculate the torque *M* on the shaft of the gate at *B* required to prevent the gate from opening in terms of *h* (the distance between salt water and fresh water levels). Plot *M* as a function of *h* for $0 \leq h \leq 6$ ft.

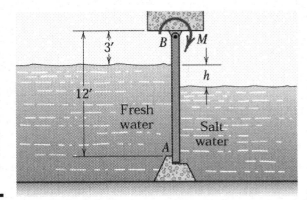

Problem Formulation

To the right is a free-body diagram for the gate. The water pressures have been replaced by linear load distributions where the maximum load intensity w is given by the general formula

$$w = \gamma d L$$

where γ is the specific weight, d is the depth of water and L is the length (into the page). Thus,

$$w_f = \gamma_f (9\,ft)(10\,ft) \qquad w_s = \gamma_s (9-h)(10)$$

where $\gamma_f = 62.4 \dfrac{lb}{ft^3}$ and $\gamma_s = 64.0 \dfrac{lb}{ft^3}$.

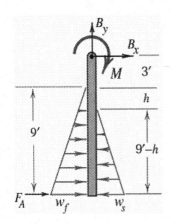

Now replace the distributed loads by their statically equivalent concentrated loads as shown on the free-body diagram to the right. Note that the contact force $F_A = 0$ since we are finding the smallest *M* required, i.e. the gate is on the verge of opening.

$$\circlearrowleft \Sigma M_B = 0 = P_f h_1 - P_s h_2 - M$$

$$M = P_f h_1 - P_s h_2$$

where,

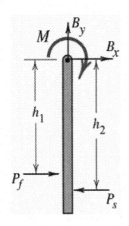

$$h_1 = 3 + \frac{2}{3}(9) = 9\,ft \qquad\qquad h_2 = 3 + h + \frac{2}{3}(9-h) = 9 + \frac{h}{3}$$

$$P_f = \frac{1}{2}w_f(9) = \frac{1}{2}\gamma_f(9)^2(10)$$

$$P_s = \frac{1}{2}w_s(9-h) = \frac{1}{2}\gamma_s(9-h)^2(10)$$

MathCad Worksheet

$\gamma_f := 62.4 \qquad\qquad \gamma_s := 64$

$w_f := \gamma_f \cdot 9 \cdot 10 \qquad\qquad w_s(h) := \gamma_s \cdot (9-h) \cdot 10$

$h_1 := 9 \qquad\qquad h_2(h) := 9 + \dfrac{h}{3}$

$P_f := \dfrac{1}{2} \cdot w_f \cdot 9 \qquad\qquad P_s(h) := \dfrac{1}{2} \cdot w_s(h) \cdot (9-h)$

$M(h) := P_f \cdot h_1 - P_s(h) \cdot h_2(h)$

h := 0 , 0.05 .. 6

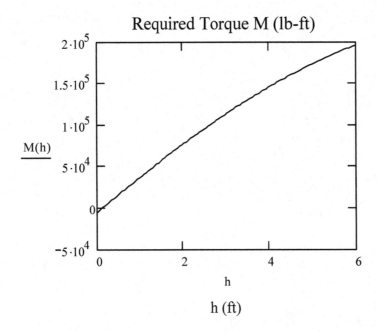

FRICTION

6

Coulomb friction (dry friction) can have a significant effect upon the analysis of engineering structures. Problem 6.1 considers three blocks stacked on an incline oriented at an arbitrary angle θ and illustrates the importance of identifying all possibilities for impending motion. In this problem there turns out to be a transition between two possibilities as the angle θ increases. Problem 6.2 considers a cylinder wedged between two rough walls. *Given...Find* is used to solve the three equilibrium equations symbolically. Problem 6.3 considers the possibility of impending slip as a person climbs to the top of a ladder and uses the symbolic *Given...Find* to solve an equation. Problems 6.4, 6.5, and 6.6 are applications involving friction on wedges, screws and flexible belts. In problem 6.4 two equations are solved symbolically using *Given...Find*. In problem 6.5 the moment required to tighten and loosen a turnbuckle are determined as functions of the friction coefficient.

6.1 Sample Problem 6/5 (Friction)

The three flat blocks are positioned on an incline that is oriented at an angle θ (instead of 30°) from the horizontal. Plot the maximum value of P (if no slipping occurs) versus θ. Consider only positive values of P and indicate the regions over which (1) the 50-kg block slides alone and (2) the 50-kg and 40-kg blocks slide together. The coefficients of friction μ_s are given in the figure to the right.

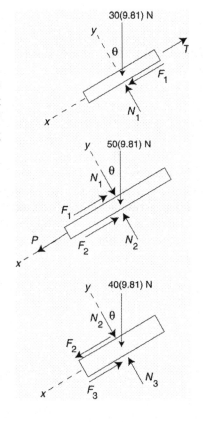

Problem Formulation

The free-body diagrams for the three blocks are shown to the right. To obtain the required plot it will be convenient to use a slightly different approach than that used in the sample problem in your text. We start by writing down the equilibrium equations without making any assumptions about where sliding occurs.

$$[\Sigma F_y = 0] \quad \text{(30-kg)} \quad N_1 - 30(9.81)\cos\theta = 0$$
$$\text{(50-kg)} \quad N_2 - N_1 - 50(9.81)\cos\theta = 0$$
$$\text{(40-kg)} \quad N_3 - N_2 - 40(9.81)\cos\theta = 0$$

These equations can be readily solved for the normal forces.

$$N_1 = 30(9.81)\cos\theta \quad N_2 = 80(9.81)\cos\theta$$
$$N_3 = 120(9.81)\cos\theta$$

$$[\Sigma F_x = 0] \quad \text{(50-kg)} \quad P - F_1 - F_2 + 50(9.81)\sin\theta = 0$$
$$\text{(40-kg)} \quad F_2 - F_3 + 40(9.81)\sin\theta = 0$$

Now we have two equations with four unknowns, P, F_1, F_2, and F_3. Note that we have not written the equation for the summation of forces in the x direction for the 30-kg block. The reason is that this equation introduces an additional unknown (T) that we are not interested in determining.

The next step is to make assumptions about which block(s) slide. As will be seen, either of the two possible assumptions about impending motion will reduce two of the friction forces to functions of θ only. This will result in two equations that may be solved for P and the remaining friction force. The forces calculated will be designated P_1 or P_2 to distinguish the two cases for impending slip.

Case (1): *Only the 50-kg block slips.*

Impending slippage at both surfaces of the 50-kg block gives $F_1 = 0.3N_1 = 88.29\cos\theta$ and $F_2 = 0.4N_1 = 313.9\cos\theta$. Substituting these results into the equilibrium equations yields

$$P_1 = 402.2\cos\theta - 490.5\sin\theta$$

Case (2): *The 40 and 50-kg blocks slide together.*

Impending slippage at the upper surface of the 40-kg block and lower surface of the 50-kg block gives $F_1 = 0.3N_1 = 88.29\cos\theta$ and $F_3 = 0.45N_3 = 529.7\cos\theta$. Substitution of these results into the equilibrium equations gives,

$$P_2 = 618.0\cos\theta - 882.9\sin\theta$$

Which of these two values of P represents the maximum load that can be applied without slippage on any surface is best illustrated by plotting the two expressions as a function of θ. This plot will be generated in the worksheet below. The basic idea is that at any specified angle θ, the critical or maximum value of P will be the smaller of two values calculated.

Mathcad Worksheet

$$P_1(\theta) := 402.2\cos(\theta) - 490.5\sin(\theta)$$

$$P_2(\theta) := 618\cdot\cos(\theta) - 882.9\sin(\theta)$$

$$\theta := 0, 0.01 .. \frac{\pi}{4}$$

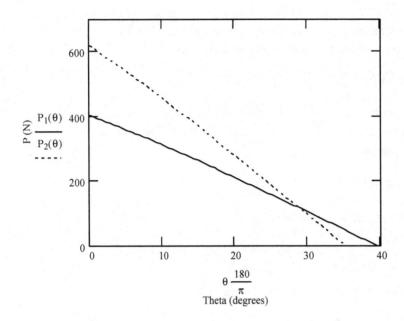

The figure above shows P_1 and P_2 plotted as a function of θ. For each θ, the critical or maximum value of P will be the smaller of two values calculated. By setting $P_1 = P_2$ we find that the two curves intersect at $\theta = 0.503$ rads (28.8°). Thus, for $\theta \leq 28.8°$ P_1 controls and the 50-kg block slides by itself while for $\theta \geq 28.8°$ P_2 controls and the 40 and 50-kg blocks slide together.

The figure below shows only the critical values for P together with an indication of the mode of slippage. This type of figure would be practically impossible to generate in Mathcad and thus was obtained using another application. It is shown here for purposes of illustration.

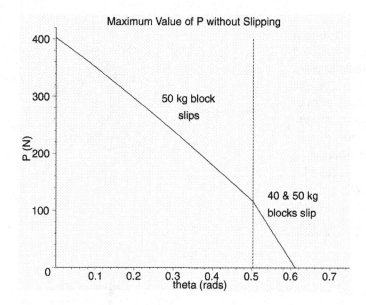

6.2 Problem 6/09 (Friction)

The 30-kg homogeneous cylinder of 400-mm diameter rests against the vertical and inclined surfaces as shown. Calculate the applied clockwise couple M which would cause the cylinder to slip. Also calculate the normal contact forces at A and B. Express your answers in terms of the angle θ and the coefficient of static friction μ_s. (a) Plot the normal forces as a function of θ $(0 \leq \theta \leq 45°)$ for $\mu = 0.3$, (b) Plot the Moment as a function of θ $(0 \leq \theta \leq 45°)$ for $\mu = 0.2, 0.5,$ and 0.8, (c) Plot the Moment as a function of μ $(0 \leq \mu \leq 1)$ for $\theta = 15, 30,$ and $45°$.

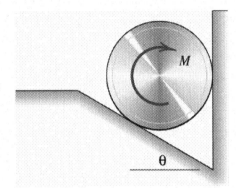

Problem Formulation

First we write the equilibrium equations from the free-body diagram for the cylinder (shown to the right).

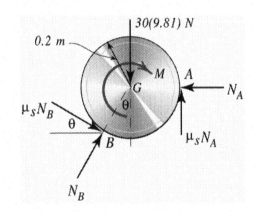

$$\circlearrowleft \Sigma M_A = 0 = M - \mu_s(N_A + N_B)0.2$$

$$\Sigma F_x = 0 = N_B \sin\theta + \mu_s N_B \cos\theta - N_A$$

$$\Sigma F_y = 0 = N_B \cos\theta - \mu_s N_B \sin\theta + \mu_s N_A - 30(9.81)$$

The last two equations can be readily solved simultaneously for N_A and N_B. Substituting these results into the first equation gives an expression that can be solved for M. The results are,

$$N_A = \frac{294.3(\sin\theta + \mu\cos\theta)}{\cos\theta(1 + \mu^2)} \qquad N_B = \frac{294.3}{\cos\theta(1 + \mu^2)}$$

$$M = \frac{58.86\mu(1 + \sin\theta + \mu\cos\theta)}{\cos\theta(1 + \mu^2)}$$

Of course, we will let MathCad solve the three equations above.

MathCad Worksheet

First, we solve the three equilibrium equations symbolically.

Given

$$M - \mu \cdot (N_A + N_B) = 0$$

$$N_B \cdot \sin(\theta) + \mu \cdot N_B \cdot \cos(\theta) - N_A = 0$$

$$N_B \cdot \cos(\theta) - \mu \cdot N_B \cdot \sin(\theta) + \mu \cdot N_A - 30 \cdot 9.81 = 0$$

$$\text{Find}(N_A, N_B, M) \rightarrow \begin{bmatrix} 294.30000000000000000 \cdot \dfrac{\sin(\theta) + \mu \cdot \cos(\theta)}{\cos(\theta) \cdot (1. + \mu^2)} \\[3mm] \dfrac{294.30000000000000000}{\cos(\theta) \cdot (1. + \mu^2)} \\[3mm] 294.30000000000000000 \cdot \mu \cdot \dfrac{\sin(\theta) + \mu \cdot \cos(\theta) + 1.}{\cos(\theta) \cdot (1. + \mu^2)} \end{bmatrix}$$

Now, copy and paste from above,

$$N_A(\mu, \theta) := 294.3 \cdot \frac{\sin(\theta) + \mu \cdot \cos(\theta)}{\cos(\theta) \cdot (1. + \mu^2)}$$

$$N_B(\mu, \theta) := \frac{294.30}{\cos(\theta) \cdot (1. + \mu^2)}$$

$$M(\mu, \theta) := 294.3 \cdot \mu \cdot \frac{\sin(\theta) + \mu \cdot \cos(\theta) + 1.}{\cos(\theta) \cdot (1. + \mu^2)}$$

$$\theta := 0, 0.01 .. 45 \cdot \frac{\pi}{180}$$

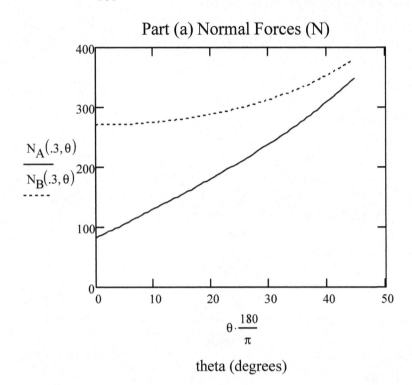

Part (a) Normal Forces (N)

$\frac{N_A(.3, \theta)}{N_B(.3, \theta)}$

$\theta \cdot \frac{180}{\pi}$

theta (degrees)

Part (b) Moment (N-m)

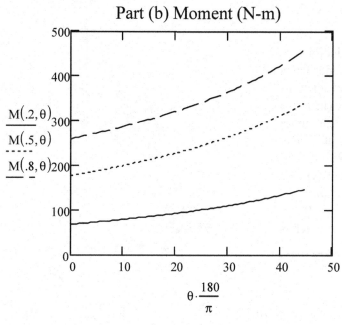

$$\theta \cdot \frac{180}{\pi}$$

theta (degrees)

$\mu := 0, 0.01 .. 1$

Part (c) Moment (N-m)

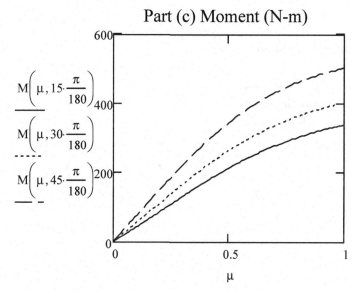

μ

Coefficient of Friction

6.3 Problem 6/44 (Friction)

The 15-kg ladder is 4-m long and is inclined at an arbitrary angle θ. The top of the ladder has a small roller, and at the ground the coefficient of friction is μ. Determine the distance s (in terms of θ and μ) to which the 90-kg painter can climb without causing the ladder to slip at its lower end A. Plot s as a function of θ ($0 \leq \theta \leq 90°$) for $\mu = 0.2$, 0.4, and 0.6. Limit the vertical (s) axis to be from 0 to 5-m. Provide a physical explanation for the angles θ where (a) $s = 0$ and (b) $s \geq 4$-m. The mass center of the painter is directly above his feet.

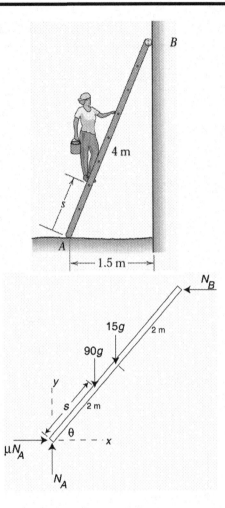

Problem Formulation

First we write the equilibrium equations from the free-body diagram for the ladder (shown to the right).

$$\Sigma F_x = 0 = \mu N_A - N_B$$

$$\Sigma F_y = 0 = N_A - 90g - 15g$$

$$\circlearrowleft \Sigma M_A = 0 = 90g(s\cos\theta) + 15g(2\cos\theta) - N_B(4\sin\theta)$$

The first two equations can be readily solved for $N_B = 105\mu g$. Substituting this result into the third equation gives an expression that can be solved for s.

$$s = \frac{14\mu\sin\theta - \cos\theta}{3\cos\theta}$$

Mathcad Worksheet

This portion of the program solves for s symbolically after NB has been substituted into the moment equilibrium equation.

Given

$$90 \cdot g \cdot s \cdot \cos(\theta) + 30 \cdot g \cdot \cos(\theta) - 105 \cdot \mu \cdot g \cdot 4 \cdot \sin(\theta) \equiv 0$$

$$\text{Find}(s) \rightarrow \frac{-1}{3} \cdot \frac{(\cos(\theta) - 14\,\mu \cdot \sin(\theta))}{\cos(\theta)}$$

$$s(\mu, \theta) := \frac{-1}{3} \cdot \frac{(\cos(\theta) - 14\,\mu \cdot \sin(\theta))}{\cos(\theta)}$$

$$\theta := 0, 0.01 .. \frac{\pi}{2}$$

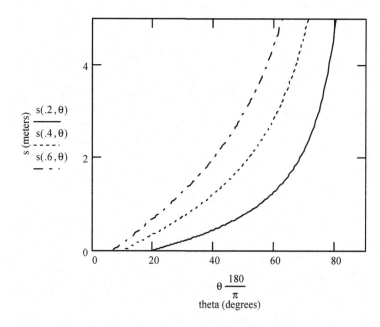

The physical interpretation is as follows. The value of θ for which $s = 0$ corresponds to the orientation where the ladder is on the verge of slipping before the painter steps on it. Obviously, the ladder will slip immediately if she does try to step on it. The values of θ for which $s \geq 4$ correspond to orientations where the painter can climb to the top of the ladder without having it slip.

6.4 Sample Problem 6/6 (Friction on Wedges)

The horizontal position of the 500-kg rectangular block of concrete is adjusted by the wedge under the action of the force **P**. Let the wedge angle be θ (instead of 5°). Also let μ_1 and μ_2 be the coefficient of static friction at the two wedge surfaces and between the block and the horizontal surface respectively. (a) Derive a general expression for P (the least force required to move the block) in terms of θ, μ_1 and μ_2. (b) For $\mu_2 = 0.6$, plot P as a function of μ_1 for $\theta = 5$, 15, and 25°. (c) For $\theta = 5°$, plot P as a function of μ_1 for $\mu_2 = 0.2, 0.4, 0.6,$ and 0.8.

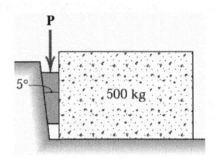

Problem Formulation

(a) First we write the equilibrium equations from the free-body diagrams for the wedge and for the block.

For the Block

$$\Sigma F_x = 0 = N_2 - \mu_2 N_3 \quad \Sigma F_y = 0 = N_3 - mg - \mu_1 N_2$$

These two equations are readily solved for N_2 and N_3.

$$N_2 = \frac{\mu_2 mg}{1 - \mu_1 \mu_2} \qquad N_3 = \frac{mg}{1 - \mu_1 \mu_2}$$

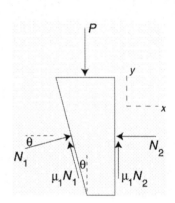

For the Wedge

$$\Sigma F_x = 0 = N_1 \cos\theta - \mu_1 N_1 \sin\theta - N_2$$

$$\Sigma F_y = 0 = N_1 \sin\theta + \mu_1 N_1 \cos\theta + \mu_1 N_2 - P$$

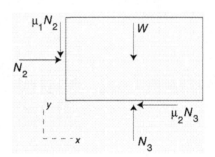

After substituting for N_2 we solve the first equation for N_1. Substituting this result into the second yields an expression for P. The result is also obtained, in somewhat different form, by symbolic algebra in the worksheet below.

$$P = \frac{\mu_2 mg\left(2\mu_1 + \left(1 - \mu_1^2\right)\tan\theta\right)}{\left(1 - \mu_1\tan\theta\right)\left(1 - \mu_1\mu_2\right)}$$

This result is used to generate the plots for parts **(b)** and **(c)** in the worksheet below.

Mathcad Worksheet

Symbolic Algebra

$$N_2 := \frac{\mu_2 \cdot m \cdot g}{1 - \mu_1 \cdot \mu_2} \qquad\qquad N_3 := \frac{m \cdot g}{1 - \mu_1 \cdot \mu_2}$$

Some of the terms in the expressions above will be in red since they are undefined. This does not affect the symbolic algebra.

Given

$$N_1 \cdot \cos(\theta) - \mu_1 \cdot N_1 \cdot \sin(\theta) - N_2 \equiv 0$$

$$N_1 \cdot \sin(\theta) + \mu_1 \cdot N_1 \cdot \cos(\theta) + \mu_1 \cdot N_2 - P \equiv 0$$

$$\text{Find}(P, N_1) \rightarrow \begin{bmatrix} -\mu_2 \cdot m \cdot g \cdot \dfrac{\left(-2 \cdot \mu_1 \cdot \cos(\theta) - \sin(\theta) + \sin(\theta) \cdot \mu_1^2\right)}{\left(-\cos(\theta) \cdot \mu_1 \cdot \mu_2 - \mu_1 \cdot \sin(\theta) + \sin(\theta) \cdot \mu_1^2 \cdot \mu_2 + \cos(\theta)\right)} \\ \mu_2 \cdot m \cdot \dfrac{g}{\left(-\cos(\theta) \cdot \mu_1 \cdot \mu_2 - \mu_1 \cdot \sin(\theta) + \sin(\theta) \cdot \mu_1^2 \cdot \mu_2 + \cos(\theta)\right)} \end{bmatrix}$$

$$m := 500 \qquad\qquad g := 9.81$$

Copying and pasting P and converting to kN.

$$P(\mu_1, \mu_2, \theta) := -\mu_2 \cdot m \cdot g \cdot \frac{\left(-2 \cdot \mu_1 \cdot \cos(\theta) - \sin(\theta) + \mu_1^2 \cdot \sin(\theta)\right)}{1000\left(\cos(\theta) - \cos(\theta) \cdot \mu_1 \cdot \mu_2 - \mu_1 \cdot \sin(\theta) + \mu_1^2 \cdot \sin(\theta) \cdot \mu_2\right)}$$

$\mu_1 := 0, .01 .. 0.8$

Part (b)

P (kN)

$\dfrac{P\left(\mu_1, .6, 5\dfrac{\pi}{180}\right)}{}$

$P\left(\mu_1, .6, 15\dfrac{\pi}{180}\right)$

$P\left(\mu_1, .6, 25\dfrac{\pi}{180}\right)$
— . —

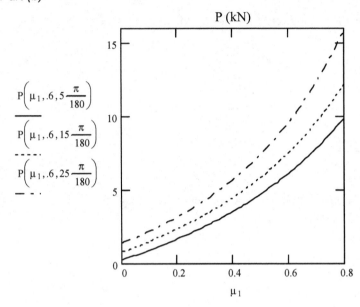

μ_1

Part (c)

P (kN)

$\dfrac{P\left(\mu_1, .2, 5\dfrac{\pi}{180}\right)}{}$

$P\left(\mu_1, .4, 5\dfrac{\pi}{180}\right)$

$P\left(\mu_1, .6, 5\dfrac{\pi}{180}\right)$
— . —

$P\left(\mu_1, .8, 5\dfrac{\pi}{180}\right)$

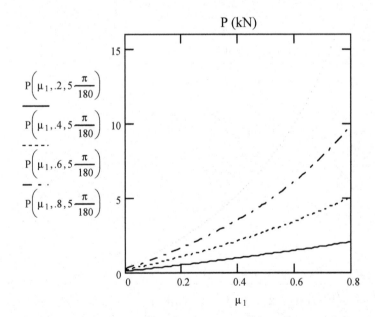

μ_1

6.5 Problem 6/60 (Friction on Screws)

The large turnbuckle supports a cable tension of 10,000 lb. The 1.25 in. screws have a mean diameter of 1.150 in. and have five square threads per inch. The coefficient of friction for the threads is μ. Both screws have single threads and are prevented from turning. Determine the moments M_T and M_L that must be applied to the body of the turnbuckle in order to tighten and loosen it respectively. Plot the moments M_T and M_L as functions of μ for $0 \leq \mu \leq 1$. Give a physical explanation for any values of M that are less than zero.

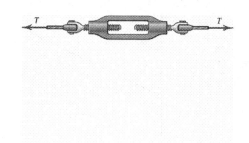

Problem Formulation

From equations 6/3 and 6/3b in your text we have,

$$M_T = 2Tr\tan(\phi + \alpha) \qquad M_L = 2Tr\tan(\phi - \alpha)$$

where T = 10,000 lb, the lead L = 1/5 in./rev. and the mean radius r = 1.15/2 = 0.575 in. Also,

$$\alpha = \tan^{-1}\frac{L}{2\pi r} \qquad \phi = \tan^{-1}\mu$$

Substitution will give M_T and M_L explicitly as a function of μ. We will let the computer substitute for us.

Mathcad Worksheet

$$T := 10000 \qquad L := \frac{1}{5} \qquad r := \frac{1.15}{2}$$

$$\alpha := \operatorname{atan}\left(\frac{L}{2 \cdot \pi \cdot r}\right) \qquad \alpha \cdot \frac{180}{\pi} = 3.169$$

$$\phi(\mu) := \operatorname{atan}(\mu)$$

$$M_T(\mu) := 2 \cdot T \cdot r \cdot \tan(\alpha + \phi(\mu))$$

$$M_L(\mu) := 2 \cdot T \cdot r \cdot \tan(\phi(\mu) - \alpha)$$

$\mu := 0, 0.005 .. 1$

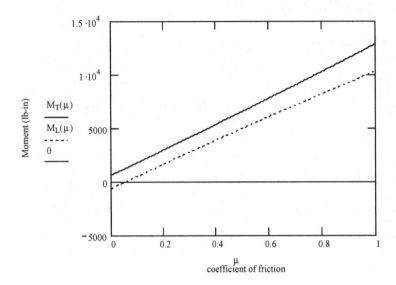

Note that the predicted values for M_L are negative for small μ. What is the physical explanation for this? You may recall that the equation for M_L derived in your text assumes that $\phi > \alpha$. When $\phi = \alpha$, the turnbuckle will be on the verge of loosening without an external moment applied to its body. The condition $\phi = \alpha$ gives, from the equations above,

$$\mu = \frac{L}{2\pi r} = \frac{1/5}{2\pi(0.575)} = 0.00553$$

This is the value of μ for which $M_L = 0$ in the plot above. We conclude then that the turnbuckle will loosen without any external moment being applied whenever $\mu < 0.00553$.

6.6 Sample Problem 6/9 (Flexible Belts)

A flexible cable which supports the 100-kg load
is passed over a circular drum and subjected to a
force P to maintain equilibrium. The coefficient
of static friction between the cable and the fixed
drum is μ. (a) For $\alpha = 0$, determine the
maximum and minimum values P may have in
order not to raise or lower the load. Plot P_{max} and
P_{min} versus μ for $0 \leq \mu \leq 1$. (b) For $P = 500$ N,
determine the minimum value which the angle α
may have before the load begins to slip. Plot
α_{min} versus μ for $0 \leq \mu \leq 1$. Limit to vertical (α)
axis to be between –60 and 360°.

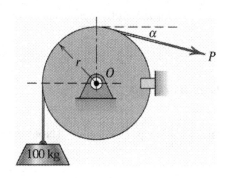

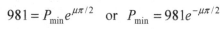

Problem Formulation

Here we will use Equation 6/7 in your text which is
repeated below for convenience.

$$T_2 = T_1 e^{\mu\beta}$$

Recall that in deriving this formula it was assumed that
$T_2 > T_1$.

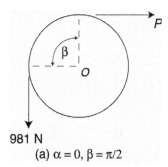

(a) $\alpha = 0$, $\beta = \pi/2$

(a) With $\alpha = 0$ the contact angle is $\beta = \pi/2$ rad. For
impending upward motion of the load we have
$T_2 = P_{max}$ and $T_1 = 981$ N. Thus

$$P_{max} = 981 e^{\mu\pi/2}$$

For impending downward motion of the load we have
$T_2 = 981$ N and $T_1 = P_{min}$.

$$981 = P_{min} e^{\mu\pi/2} \quad \text{or} \quad P_{min} = 981 e^{-\mu\pi/2}$$

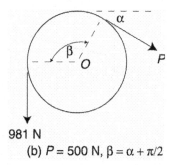

(b) $P = 500$ N, $\beta = \alpha + \pi/2$

(b) With $P = 500$ N we have $\beta = \pi/2 + \alpha$, $T_2 = 981$ N and $T_1 = P = 500$ N. From
Equation 6/7 we have,

$$981/500 = e^{\mu(\pi/2+\alpha)}$$

Taking the natural log of both sides of the above equation and solving for α gives,

$$\alpha = \frac{\ln(981/500)}{\mu} - \frac{\pi}{2}$$

Mathcad Worksheet

$$P_{max}(\mu) := \frac{981}{1000} \cdot \exp\left(\frac{\mu \cdot \pi}{2}\right) \qquad \text{(Note the conversion to kN)}$$

$$P_{min}(\mu) := \frac{981}{1000} \cdot \exp\left(\frac{-\mu \cdot \pi}{2}\right)$$

$$\mu := 0, 0.005 .. 1$$

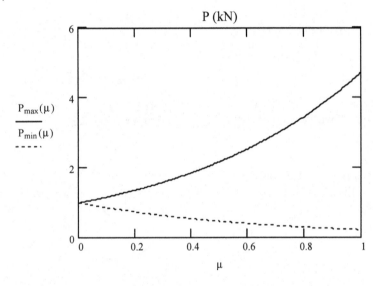

$$\alpha(\mu) := \left(\frac{\ln\left(\dfrac{981}{500}\right)}{\mu} - \frac{\pi}{2} \right) \cdot \frac{180}{\pi} \quad \text{(Note the conversion from radians to degrees)}$$

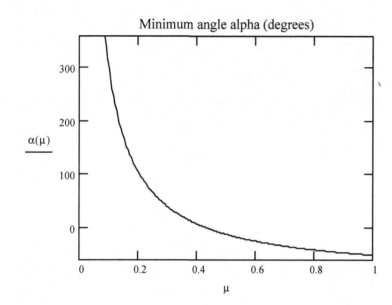

VIRTUAL WORK

7

This chapter considers the application of the principle of virtual work to the equilibrium and stability analysis of engineering structures. Problem 6.1 looks at the compressive force developed in a hydraulic cylinder that is part of a mechanism for elevating automobiles. The results of a parametric study suggest an obvious improvement in the design of the system. The formulation of problem 7.2 results in an equation that cannot be solved exactly. When this occurs one generally has to make a choice between a graphical or numerical solution. This problem illustrates in some detail a general graphical approach that is very useful in some situations. For purposes of comparison, a numerical solution is also obtained with Mathcad's *root* function. Problem 7.3 takes a look at a problem with multiple equilibrium states and evaluates these in terms of their stability. The location of a minimum is found by setting a derivative equal to zero. The resulting equation is solved numerically using *solve*.

7.1 Problem 7/30 (Virtual Work, Equilibrium)

Express the compression C in the hydraulic cylinder of the car hoist in terms of the angle θ. The mass of the hoist is negligible compared to the mass m of the vehicle. (a) Plot the non-dimensional compression C/mg as a function of θ for three non-dimensional length ratios; $b/L = 0.1$, 0.5, and 0.9. Let θ range between 0 and 90° and limit the vertical scale to between 0 and 3. (b) Plot C/mg as a function of b/L for several values of θ letting b/L vary between 0 and 2. From your results, see if you can improve the design of the hoist system by re-positioning the hydraulic cylinder.

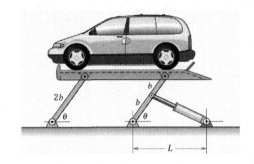

Problem Formulation

The active force diagram for the system is shown to the right. Let the length of the cylinder (AB in the diagram) be a so that the virtual work done by C is $C\delta a$. The principle of virtual work applied to this system gives,

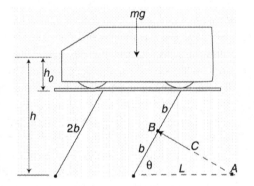

$$\delta U = 0 = C\delta a - mg\delta h$$

Now we have to do some geometry to relate the virtual displacements δa and δh to the angle θ. From the diagram we can write a^2 as follows.

$$a^2 = (b\sin\theta)^2 + (L - b\cos\theta)^2$$

Note that it is not actually necessary to solve for a in order to determine the virtual displacement δa. All you need is an expression relating a and θ. In the present case it is somewhat easier to find the variation of the expression on the right hand side of the equation above than it would be to find the variation of the square root of that expression. Taking the variation of the above equation yields,

$$2a\delta a = 2b\sin\theta(b\cos\theta\delta\theta) + 2(L - b\cos\theta)(b\sin\theta\delta\theta)$$

Solving for δa and simplifying yields,

$$\delta a = \frac{Lb}{a}\sin\theta\delta\theta$$

where $\quad a = \sqrt{(b\sin\theta)^2 + (L - b\cos\theta)^2} = L\sqrt{1 + \left(\frac{b}{L}\right)^2 - 2\frac{b}{L}\cos\theta}$

Now we need to find a relationship between δh and θ.

$$h = 2b\sin\theta + h_0 \qquad\qquad \delta h = 2b\cos\theta\delta\theta$$

Now we substitute the virtual displacements δa and δh into the virtual work equation above.

$$C\left(\frac{Lb}{a}\sin\theta\right)\delta\theta - mg(2b\cos\theta)\delta\theta = 0$$

$$C = \frac{2mga}{L}\cot\theta$$

Solving for the non-dimensional compression C/mg and substituting for a yields,

$$\frac{C}{mg} = 2\cot\theta\sqrt{1 + \left(\frac{b}{L}\right)^2 - 2\frac{b}{L}\cos\theta}$$

Before generating the required plots it might be useful to introduce some non-dimensional parameters. Letting $C' = C/mg$ and $\eta = b/L$ we can re-write the above equation as

$$C' = 2\cot\theta\sqrt{1 + \eta^2 - 2\eta\cos\theta}$$

Now we need to plot C' as a function of θ for $\eta = 0.1, 0.5,$ and 0.9.

Mathcad Worksheet

$$C(\eta, \theta) := 2 \cdot \cot(\theta) \cdot \sqrt{1 + \eta^2 - 2 \cdot \eta \cdot \cos(\theta)}$$

Part (a)

$$\theta := 0, 0.01 .. \frac{\pi}{2}$$

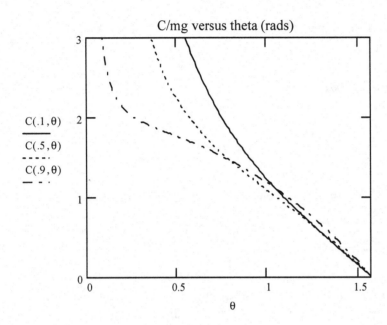

Part (b)

$\eta := 0, .01..2$

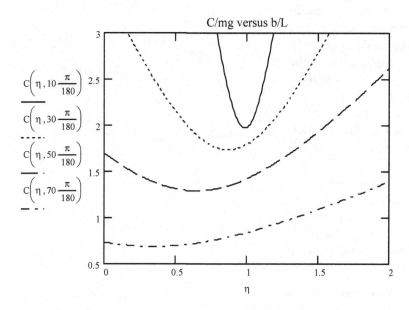

C/mg versus b/L

$C\left(\eta, 10\frac{\pi}{180}\right)$ ——

$C\left(\eta, 30\frac{\pi}{180}\right)$ - - - -

$C\left(\eta, 50\frac{\pi}{180}\right)$ — · —

$C\left(\eta, 70\frac{\pi}{180}\right)$ — · · —

η

The first diagram (part (a)) shows that the present design has some real problems for small angles θ. To appreciate this, remember that we are plotting the non-dimensional force. If the non-dimensional force equals 3, for example, the compressive force in the cylinder will be three times the weight of the vehicle. The main intent of the diagram for part (b) is to show that changing the design by fine-tuning b/L will not really remove the real problem.

So, what is the physical reason for the compressive force approaching infinity as θ approaches zero? Look again at the active force diagram and try to visualize how the orientation of C changes as θ gets smaller. As θ gets smaller, the vertical component of C gets smaller, making it harder and harder for the cylinder to support the weight of the vehicle. Without changing the overall design too much, the most obvious thing to do is to recess the cylinder somewhat so that it never approaches a horizontal orientation. In other words, move point A (the base of the cylinder) vertically downward. This is illustrated by the modified drawing to the right. Note that, with this arrangement, the vertical component of C will not approach zero at small θ.

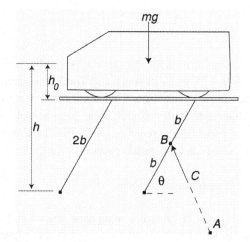

7.2 Sample Problem 7/5 (Potential Energy)

The two uniform links, each of mass m, are in the vertical plane and are connected and constrained as shown. As the angle θ between the links increases with the application of the horizontal force P, the light rod, which is connected at A and passes through a pivoted collar at B, compresses the spring of stiffness k. If the spring is uncompressed in the position where $\theta = 0$, determine the force P which will produce equilibrium at the angle θ. (a) Develop a graphical approach for determining the equilibrium value of θ corresponding to a given force P. Use a computer to generate one or more plots that illustrate your approach. (b) Obtain a numerical solution for the equilibrium angle θ for the case $P = 100$ lb, $mg = 50$ lb, $k = 20$ lb/in, and $b = 5$ in.

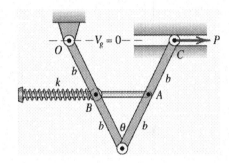

Problem Formulation

The first part of this problem (determining the force P which will produce equilibrium at the angle θ) is identical to the sample problem in your text. Therefore, we start with the equation obtained in the sample problem. Be sure that you understand how this equation was derived.

$$P = kb \sin \frac{\theta}{2} + \frac{1}{2} mg \tan \frac{\theta}{2}$$

This equation is ideally suited to situations where you would like to find the force P that will result in some specified equilibrium angle θ. Here we are interested in the opposite situation. Given a force P we would like to calculate the equilibrium value of θ. Unfortunately, the equation above cannot be inverted analytically to give θ explicitly as a function of P. In situations like this you should try either a graphical or numerical solution.

(a) Graphical Approach

Here we will illustrate two graphical approaches. The first is suggested at the end of the sample problem in your text and is suitable for solving specific problems. The second approach is more general and is suitable for certain types of design situations.

(a1) *Specific Graphical Approach*

The specific approach is suitable for situations where P, b, m and k are all specified. For example, consider the case where you want the equilibrium angle θ for only one situation. For convenience, we'll take the specific case defined in part (b).

$$P = 100 \text{ lb}, mg = 50 \text{ lb}, k = 20 \text{ lb/in, and } b = 5 \text{ in.}$$

Here we forget for the moment that P is given and plot, from the expression above, P as a function of θ. Superimposed on this plot would be a horizontal line at $P = 100$. This horizontal line can either be drawn at the same time the plot is generated or by hand after the plot has been printed. After the plot has been printed you would next find the intersection between the horizontal line and the curve. Since the curve represents the solution for all θ, the intersection will be the solution for the specific case where $P = 100$ lb. The equilibrium value of θ can thus be found by dropping a vertical line from the point of intersection to the θ axis. The procedure is illustrated in the figure to the right. From this plot we can estimate the equilibrium angle to be about $\theta = 95°$. For purposes of comparison, the numerical approach in part (b) gives $\theta = 94.05°$.

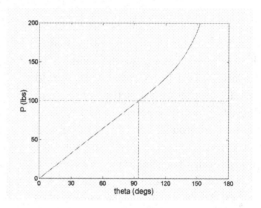

(a2) *General Graphical Approach*

What we have in mind here are certain types of design situations where specific values of parameters have not yet been decided. Here we want to develop a graphical approach that would not require us to run to a computer each time a specific case is considered. This is what is meant by a general graphical approach. The effectiveness and generality of graphical solutions can be significantly increased by first reducing, as much as possible, the number of parameters that appear in the equation. In the present case we can reduce the number of parameters from three (k, b, and m) to one by dividing both sides of the equation by the weight mg. Since P, kb and mg all have units of force, this operation results in a non-dimensional equation relating a non-dimensional force $P' = P/mg$ to the equilibrium angle θ. The only parameter remaining is the non-dimensional quantity $kb/mg = \beta$.

$$P' = \frac{P}{mg} = \beta \sin\frac{\theta}{2} + \frac{1}{2}\tan\frac{\theta}{2} \quad \text{where } \beta = \frac{kb}{mg}.$$

The plot to the right shows P' versus θ for several, evenly spaced, values of β. For a specific problem you would be asked to find θ given P, k, b, and m. One would first determine P' and β and then construct a horizontal line at the calculated value of P' and find where that line intersects a curve at the appropriate value of β. Dropping a vertical line from this intersection gives the angle θ.

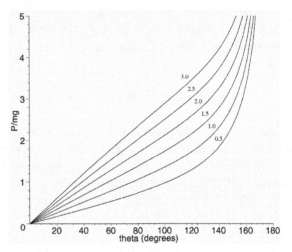

Of course, if the graph is prepared in advance, as proposed here, it probably will not have a curve for the value of β calculated. In this case you could generate another curve at the appropriate value of β, but this defeats our purpose. Another approach is to obtain an approximate answer by means of interpolation. To illustrate, consider a case where $P = 220$ lb, $mg = 100$ lb, $k = 29$ lb/in and $b = 6$ in. These values give $P' = 2.2$ and $\beta = 1.74$. The second plot to the right illustrates the interpolation procedure. Start with the horizontal line at $P' = 2.2$ and then follow this line to where it reaches a point about halfway between the curves for $\beta = 1.5$ and 2. Dropping a vertical line from this point gives a value for the equilibrium angle θ of about 112°.

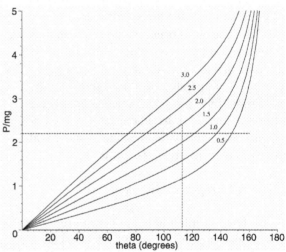

The worksheet below will show how to generate plots such as those shown above. Of course, the construction of vertical lines and interpolation are operations that are performed after the curves are generated and printed.

(b) The numerical solution is given in the worksheet below.

Mathcad Worksheet

Plot for part (a1)

$mg := 50 \qquad k := 20 \qquad b := 5$

$$P(\theta) := k \cdot b \cdot \sin\left(\frac{\theta}{2}\right) + \frac{1}{2} \cdot mg \cdot \tan\left(\frac{\theta}{2}\right)$$

$\theta := 0, 0.05 .. \pi$

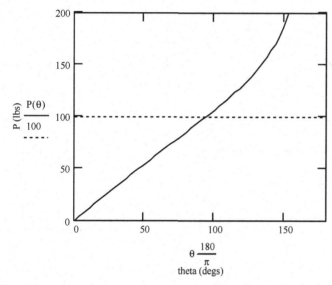

Part (b)

It is convenient to do part (b) here since it involves the parameters already defined above. One of the points here is, of course, that it is very difficult to beat the numerical approach for accuracy and ease of application. Here we will use the root command. Remember that root finds the zeros of an equation. Thus, rather than writing P = 100 we write P - 100. The = 0 is implied. Since P is a function, we can conveniently switch our variables from theta to x.

$x := 0.5$ Remember that root requires an initial guess.

Also note the conversion to degrees in the following.

$$\frac{180}{\pi} \text{root}(P(x) - 100, x) = 94.052$$

Plot for part (a2)

$$P_p(\beta, \theta) := \beta \cdot \sin\left(\frac{\theta}{2}\right) + \frac{1}{2} \cdot \tan\left(\frac{\theta}{2}\right)$$

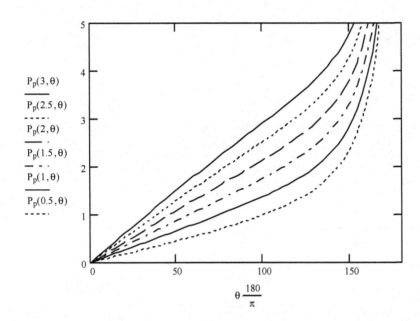

$P_p(3,\theta)$

$P_p(2.5,\theta)$

$P_p(2,\theta)$

$P_p(1.5,\theta)$

$P_p(1,\theta)$

$P_p(0.5,\theta)$

$\theta \dfrac{180}{\pi}$

7.3 Problem 7/79 (Potential Energy and Stability)

The uniform link OA has a mass of 20 kg and is supported in the vertical plane by the spring AB whose unstretched length is 400 mm. Plot the total potential energy V as a function of θ from $\theta = -15°$ to $\theta = 120°$. Consider two cases, $k = 100$ and 900 N/m. Determine all equilibrium positions (angles θ) and their stability for each case. Take $Vg = 0$ on a level through O.

Problem Formulation

Imagine dividing triangle OAB into two identical right triangles to note that half of the length AB is $0.4\sin((180 - \theta)/2)$. Also note the trig identity $\sin((180 - \theta)/2) = \cos(\theta/2)$. Thus,

$$AB = 2(0.4)\sin\left(\frac{180 - \theta}{2}\right) = 0.8\cos\frac{\theta}{2}$$

Since the unstretched length of the spring is 0.4 m, the elastic potential energy for the springs is,

$$V_e = \frac{1}{2}k(AB - 0.4)^2 = \frac{1}{2}k\left(0.8\cos\frac{\theta}{2} - .4\right)^2$$

Taking $V_g = 0$ on the horizontal plane through O we obtain the gravitational potential energy of the system as,

$$V_g = 20(9.81)(-0.2\cos\theta) = -39.24\cos\theta$$

Thus, the total potential energy and its derivative are,

$$V = V_g + V_e = \frac{1}{2}k\left(0.8\cos\frac{\theta}{2} - 0.4\right)^2 - 39.24\cos\theta$$

$$\frac{dV}{d\theta} = 39.24\sin\theta - 0.4k\left(0.8\cos\frac{\theta}{2} - 0.4\right)\sin\frac{\theta}{2}$$

The equilibrium positions and their stability will be determined below.

MathCad Worksheet

$$AB(\theta) := 0.8 \cdot \cos\left(\frac{\theta}{2}\right)$$

$$V_e(\theta,k) := \frac{1}{2} \cdot k \cdot (AB(\theta) - 0.4)^2$$

$$V_g(\theta) := -39.24 \cdot \cos(\theta)$$

$$V(\theta,k) := V_e(\theta,k) + V_g(\theta)$$

$$theta := -15, -14.9 .. 120$$

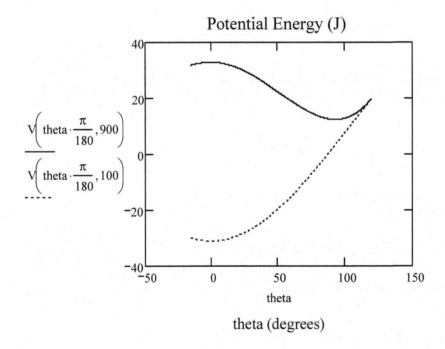

Potential Energy (J)

$$\frac{V\left(theta \cdot \dfrac{\pi}{180}, 900\right)}{V\left(theta \cdot \dfrac{\pi}{180}, 100\right)}$$

theta

theta (degrees)

Equilibrium positions and their stability.

Recall from your text that stable equilibrium occurs at positions where the total potential energy of the system is a minimum. Unstable equilibrium, on the other hand, occurs at positions where the total potential energy of the system is a maximum. From this general principle we see that the spring with stiffness 100 N/m has only one equilibrium position at $\theta = 0$ and it is stable. The spring with stiffness 900 N/m has two equilibrium positions, an unstable one at $\theta = 0$ and a stable one just beyond 1.5 radians. This second position can be determined more precisely by finding the angles for which $dV/d\theta = 0$. This is carried out in the following.

$$dV(\theta, k) := \frac{d}{d\theta} V(\theta, k)$$

$$dV(x, 900) \text{ solve}, x \rightarrow \begin{pmatrix} 0 \\ 6.2831853071795864769 \\ -1.6261025608933934752 \\ 1.6261025608933934752 \end{pmatrix}$$

MathCad has found four solutions, but only the first and fourth are in our range. The first solution (0) is unstable while the fourth (1.626 radians) is stable

Summary:

$k = 100$ N/m

$\qquad \theta = 0$ Stable

$k = 900$ N/m

$\qquad \theta = 0$ Unstable

$\qquad \theta = 1.626$ rads (93.18°)   Stable